MW00910824

SHERIDAN'S
NEW JERSEY
ELECTRICAL CONTRACTOR'S
BUSINESS LAW REVIEW

18TH
Edition

Author
Robert J. Sheridan

ISBN-13: 978-0-9794616-0-6

TABLE OF CONTENTS

INTRODUCTION

Obtaining a New Jersey electrical contractor's license shall open the door to potential rewarding opportunities within the electrical construction industry. This license signifies that you are qualified and have demonstrated your ability to meet the minimum standards in experience and education for the installation and test of electrical systems, and electrical construction business management. With this license you will have the opportunity to operate your own electrical contracting business or potentially expanded your career into project management / supervisory positions within the electrical construction industry.

Currently the New Jersey electrical contractors licensing examination consist of a three-part board examination requiring an applicant to obtain a passing grade (70%) in each part. The three areas that the applicant is tested in are 1. Practical Trade , 2. Business Law, 3. Alarm Code. The Business Law exam is a two-hour, 50 question test.

The goal of this book is to assist you in preparing for the business law portion of the New Jersey electrical contractors licensing examination. This book is a supplement to the NASCLA Contractors Guide NJ Edition, which is the business law textbook written by the testing organization and approved by the New Jersey Board of Examiners. The NASCLA Contractors Guide NJ Edition contains the basic information required for the business law exam. However, this text does not provide any review questions or student practice tests. This is where the *New Jersey Electrical Contractor's Business Law Review* fills in. This study guide is divided into equivalent chapters as in the reference manual and will provide a series of review questions on each section. Detailed answers with references back to the NASCLA Contractors Guide NJ Edition are provided. In addition, this study guide also provides two student practice exams for the business law exam along with detailed answers and references.

To adequately prepare for this exam the following study method is recommended. I have taught this subject matter since the inception of the business law exam in 1990 using this method and have had approximately 98% of my students pass the business law exam the first time. There is no magic method or class that will miraculously instill this knowledge in you overnight or with little effort on your part. You will need to put aside three hours a week for three months. In this busy world, an expectation of anymore time per week than this is probably a plan to fail. Find a quiet place, somewhere away from home with little or no distractions during this study period. Plan to complete one section or one practice exam each study period. Now let us discuss what you must do during these study periods.

The state test allows the students to bring with them, into this exam, their copy of the NASCLA Contractors Guide NJ Edition only, the *New Jersey Electrical Contractor's Business Law Review* is not allowed into the test. This is an important point, your marked up and tabbed NASCLA Contractors Guide NJ Edition is allowed to be used for the test, therefore generous high-lighting is recommended. I recommended that you tab each chapter and Appendix A: Glossary with one color tab. Then tab Appendix E, I and J with another color tab. In addition, you should use two colors for high lighting, one color for business law/electrical contracting and another color for the alarm licensing laws. This is due to the laws between electrical and alarm licensing are similar but do contain differences and can easily be mixed up during the test.

During each study period the student should first read each section in the NASCLA Contractors Guide NJ Edition and then complete each section's review questions in the *New Jersey Electrical Contractor's Business Law Review*. While completing the review questions the student should identify the answer in the textbook and highlight the sentence or paragraph. Once all the sections are completed the student should take the practice exam. The student should attempt to complete these practice exams in the time allowed. Any unanswered question should then be worked on prior to reviewing the answers. REMEMBER just going to the answer first without aggressively trying to find or work out the answers will not teach you the material or how to take a test.

Time is an important factor when taking any of the Board Exams. For the Business Law exam, the student has 2.4 minutes on average to complete each question. This is adequate time for the student to review the reference manual prior to answering any question. Therefore, do not try to memorize the whole NASCLA Contractors Guide NJ Edition but thoroughly know what type of information is in each section. For example; "What is the social security withheld on an employee's weekly gross pay of $1000?", go to the Tax Law Section. This method will allow you to quickly flip to the correct section during the test and the highlighted areas will keep you focused on the primary information. Most of the sections are not many pages and with the sub-section titles you can find the paragraph containing the information you need to answer the question quickly. As you answer the review questions keep an eye on the time it takes you to find an answer and learn to skip over questions and go on to the questions you may know.

Last point on test taking and studying I would like to state. You must also study how to take a test. Many people fail not because they do not know the material but are not prepare to take a standardized test. You only get credit for answers marked on the answer sheet. You do not lose any more points for guessing therefore if you are not sure of an answer, GUESS. Also do the test in order. Do not skip around. With computerized testing you can identify each question when you select an answer as to whether you feel it is ready to be MARK'ed. Use this to identify questions that may need to go back to review (ready for marking not selected). The computer will roll thru this set prior to allowing you to submit for grading. Remember, finish the exam by keep moving through the problems. If you do not complete the test, you are guaranteed to be marked incorrect for those questions not finished.

In using the *New Jersey Electrical Contractor's Business Law Review* you will find, as have many of my students in the past, that this portion of the Electrical Contractor's Board Examination to be relatively easy.

Chapter 1 Business Plan

Main Topics for this Section
>   A. Business Plans
>   B. Elements of a Business Plan
>   C. Business Plan Pitfalls

The chapter reviews Chapter 1: The Plan of the NASCLA Contractors Guide to Business, Law, and Project Management New Jersey Edition.

Supplementary Notes

>   This section provides a good background to developing, and benefits of a business plan.

Business Plan Benefits

1. What is it that must be developed to start a successful business that will define how you will meet your primary business goals?
   A) Financial Plan　　　B) Goals　　　　　　　C) Business Plan　　　D) Business Organization

2. Which of the following is not a key function of a business plan?
   A) Financial Books　　B) Planning Tool　　　C) Loan Document　　D) Benchmarking Tool

3. A business plan should be change or modified _____.
   A) yearly
   B) as business and markets change
   C) quarterly
   D) never

Elements of a Business Plan

4. Which element of a business plan contains the company vision and mission, legal structure, management personnel, business location and facilities?
   A) Executive summary　B) Company summary　C) Market analysis　　D) Financial plan

5. Which element of a business plan contains the uniqueness of your product or service as well as your pricing, adverting and promotional strategies?
   A) Product & Services　B) Company summary　C) Market analysis　　D) Market strategy

6. Which element of a business plan defines your target market and trends?
   A) Market strategy　　B) Company summary　C) Market analysis　　D) Financial plan

Chapter 2 Business Organization

Main Topics for this Section
    A. Business Structure
    B. Naming a business
    C. NJ Division of Revenue

The chapter reviews Chapter 2: Choosing Your Business Structure of the NASCLA Contractors Guide to Business, Law, and Project Management New Jersey Edition.

Supplementary Notes

This section covers the basics of business organization types with their advantages and disadvantages. The student should be aware of Appendix A: Glossary. This contains definitions of many items called out within the text book main sections.

The following is supplemental information on business organization related to operating a business in New Jersey.

Once a business completes the form NJ-REG it is registered for tax purposes and is issued a Federal Employer Identification Number (FEIN).

Domestic corporations are entities registered/ incorporated within the state, where as foreign corporations are entities doing business in one state but incorporated in another state.

Business Ownership Types

1. What is the simplest and least expensive business type to setup and operate?
   A) LLC
   B) Partnership
   C) Individual Unregistered
   D) Sole Proprietorship

2. Profits earned from revenue in a sole proprietorship are taxed as_____.
   A) Personal income   B) S-Corp income   C) Corporate income   D) Business income

3. Upon death of the owner a Sole Proprietorship business will _____.
   A) be inherited by relatives   B) terminate   C) continue as is   D) be sold

4. To register a sole proprietorship in the State of New Jersey an applicant must file a _____.
   A) Public Records form   B) NJ-REG   C) Form 2553   D) DBA filing

5. General Partners, in a Limited Partnership, assume a _____ liability to creditors.
   A) unlimited   B) limited to investment   C) no   D) equal

6. In a limited partnership where the profit distribution is defined as a 60%-40% split between the two partners, if the partnership incurs a $200,000 obligation what is the liability of the minor partner (40% investor)?
   A) $80,000   B) $200,000   C) $120,000   D) None

7. Business income and sales taxes are the responsibility of _____.
   A) General partners only
   B) limited for Limited partners
   C) all partners
   D) stock holders

8. Partnerships shall report federal income taxes on form _____.
   A) 1040B   B) 6253   C) 2553   D) 1065

9. For Limited and Limited Liability partnerships to register with the State of New Jersey, applicants must first file a _____.
   A) Public Records filing   B) NJ-REG   C) Articles of Partnership   D) DBA

10. Partnerships must file a form NJ-REG within _____ days after the Public Records filing.
    A) 10   B) 15   C) 60   D) 90

11. Profits paid from a corporation to shareholders are called _____.
    A) capital recovery   B) re-investment   C) profit payments   D) dividends

12. Which of the following is **NOT** a required document for a C Corporation formation?
    A) Articles of Incorporation
    B) Corporation By-laws
    C) DBA
    D) None of the above

13. Which type of business organization does **NOT** allow for avoidance of double taxation on earnings?
A) S-Corporation   B) LLC   C) C-Corporation   D) B&C

14. An S-Corporation can have **NO** more than ____ shareholders.
A) 1   B) 75   C) 12   D) 100

15. LLC's are taxed based on _____.
A) the business entity
B) the number of business owners
C) a sole proprietorship
D) DBA filing

16. What type of business organization provides limited liability for the partners and the advantage of single taxation at the partner's personal income tax rate?
A) General Partnership
B) General Corporation
C) Limited Liability Company
D) Limited Partnership

17. What type of business organization is an artificial entity and allows the shareholders to choose how their profits are to be taxed, corporate or individual tax rate?
A) Corporation   B) Sole Proprietorship   C) Limited Partnership   D) "S" Corporation

18. A Corporation doing business in the state in which it was chartered is called a _____corporation.
A) Foreign   B) domestic   C) local   D) regional

19. A Corporation doing business in another state in which it was chartered is called a _____ corporation by that state.
A) Foreign   B) domestic   C) intra-state   D) inter-state

Business Naming

20. A business's registered name is also known as _____ .
A) DBA name
B) trade name
C) fictitious name
D) all the above

21. Sole proprietorships operating under a fictitious are required to register with _____ within the county the business is located.
A) County Public Records
B) County Clerk's office
C) NJ Division of Revenue
D) all the above

22. After a business's registration is filed with the State Treasurer's office a domestic business name will be reserved for _____.
A) 120 days   B) 90 days   C) 60 days   D) 30 days

23. After a business's registration is filed with the State Treasurer's office a foreign business name will be reserved for _____.
    A) 120 days
    B) 90 days
    C) 360 days
    D) end of calendar year

24. Domestic and foreign corporations are required to _____ before filing Articles of Incorporation.
    A) register with NJ Division of Revenue
    B) reserve a corporation name
    C) file a Form 2553
    D) file for distribution of dividends

Chapter 3 Licensing

Main Topics for this Section
A. General Licensing Rules and Regulations for Electrical Contractors
B. Appendix E: State Board of Examiners of Electrical Contractors Law and Regulations
C. Appendix H: Fire Alarm Advisory Committee Law and Regulations
D. Appendix I: NJ Uniform Enforcement Act and Regulations

The chapter reviews Chapter 3: Becoming a Licensed Contractor of the NASCLA Contractors Guide to Business, Law, and Project Management New Jersey Edition.

Supplementary Notes

This section shall review Chapter 3 Becoming a Licensed Contractor an Appendix E: State Board of Examiners of Electrical Contractors Laws and Regulations, Appendix H: Fire Alarm Advisory Committee Law and Regulations and Appendix I: NJ Uniform Enforcement Act and Regulations from the NASCLA Contractors Guide.

The following identifies some the basic requirements for licensing and engaging in electrical contracting within New Jersey that are not explicitly stated within the reference manual.
- An individual who passes the complete state examination will receive a letter indicating they have successfully qualified for an electrical contractor's license.
- The individual will then have to apply for their license after,
- Obtaining business insurance for a minimum of $300,000 general liability for property damage and bodily injury to or death of one or more persons in any one accident or occurrence.
- Obtaining a surety bond in favor of the State of New Jersey in the sum of $1000.00.
- The individual will, after obtaining a license, apply for a business permit. This can be done at the same time as the application for the license. The pressure seal will be issued only after the business permit is issued. Remember, you can be licensed but unless you hold a business permit you cannot engage in electrical contracting within the State of New Jersey. You are not required to obtain a business permit. To apply for the business permit the individual must obtain the following in addition to the requirements for the license stated above;
- Federal Tax Identification number
- Sales Tax Certificate of Authority

Appendix E: Electrical Contractors Licensing Act

1. Who appoints the members of the N. J. Board of Examiners?
   A) Senate   B) Governor   C) Dept. of Law & Public Safety   D) Community

2. How many electrical contractors sit on the N. J. Board of Examiners?
   A) 3   B) 7   C) 2   D) 5

3. How many public members not associated with the electrical industry sit on the N. J. Board of Examiners?
   A) 3   B) 7   C) 2   D) 1

4. How long is an appointment on the N. J. Board of Examiners?
   A) 1 yr.   B) 3 yrs   C) 5 yrs   D) 4 yrs

5. Every person who holds a business permit, _____ insurance for combined property damage and bodily injury to or death of one or more persons in any one accident is required.
   A) $1000 Surety Bond B) $300,000 general liability C) $1000 general Liability D) no

6. Cancellation of an Electrical Contractor's general liability insurance shall not take effect until at least _____ days' notice of intent has been received in writing by the board.
   A) 30   B) 45   C) 14   D) 10

7. The Board may require a licensee to be re-examined upon failure to apply for a license renewal within _____ of the expiration date.
   A) 30 days   B) 90 days   C) 6 months   D) 4 months

8. A person can obtain an electrical seal after obtaining......
   A) license   B) business permit   C) liability insurance   D) surety bond

9. A licensed electrical contractor must receive _____ hours of continuing education for each triennial registration period.
   A) 1   B) 10   C) 24   D) 34

10. A licensed electrical contractor must receive, at a minimum, _____ hours of continuing education related to the most recent edition of the NFPA-70 for each triennial registration period.
    A) 9   B) 10   C) 24   D) 34

11. How many CE credits can be carried over to the next triennial license period?
    A) 10   B) 8   C) 3   D) none

12. If the licensee is rendered incapable of fulfilling his/her duties due to death or illness, the business entity may continue to engage in electrical work for _____.
    A) 6 months   B) 3 months   C) 30 days   D) 1 yr

13. Which of the following is **NOT** exempt electrical work.
    A) Replacement of a 277V lamp
    B) Replacement of a 120V fuses
    C) Installation of Transmission Lines under contract for the local public utility
    D) Employed qualified journeyman electrician for the school district installing 120V light fixtures

---

14. Which of the following is **NOT** exempt work?
A) Replacement of fuses within a 1ϕ, 120V disconnect
B) Electrical work in Public Utilities for power distribution
C) Electrical work on radio transmission equipment
D) Installation of 120V/ 9V battery back-up smoke alarms

15. A licensee must obtain a surety bond in the favor of the State of N. J. in the amount of ____.
A) $10000.00   B) $1000.00   C) $300,000.00   D) $1,000,000.00

Subchapter 1: General Rules and Regulations for Electrical Contractors

16. On a commercial vehicle utilized in the practice of electrical contracting the required identification information shall visible mark on the vehicle with lettering at least_____ height.
A) 2 in   B) 3 in   C) 4 in   D) 5 in

17. When available space for identification lettering is limited by design of the vehicle and making strict compliance with marking letter size, the lettering size shall be reduced to _____.
A) 1 in
B) 2 in
C) 2 1/2 in
D) as close as possible to 3 in

18. Which of the following is **NOT** required on all electrical contracting business correspondences and stationary?
A) License number
B) Business Permit number
C) Business address
D) Name of the Electrical Contractor

19. Which of the following is required to engage in an electrical contracting business?
A) B, C & D   B) Surety Bond   C) Liability Insurance   D) Business

20. How many credits of continuing education is required within the preceding triennial period?
A) 10   B) 24   C) 34   D) 44

21. If during a triennial period a licensee completes more than the minimum continuing education requirements, how many credits may they carry forward to the next triennial period?
A) 3   B) 8   C) 12   D) all

22. Of all continuing education credits required, what is the minimum number required that must be related to the newest National Electrical code?
A) 10   B) 9   C) 24   D) 12

23. Of all continuing education credits required, what is the minimum number required that must be related to the applicable State statutes and rules?
A) 1   B) 3   C) 6   D) 9

24. Which of the following is **NOT** a method of obtain approved continuing education credits?
    A) CE class pending Board approval
    B) Authorship of an 8000 word textbook related to the practice of electrical contracting
    C) Authorship of a published article relate to electrical contracting with only 250 words in length
    D) All of the above

25. The Board shall grant no more than ____CE credits, for participation in instructional activities, in a triennial period?
    A) 3   B) 9   C) 15   D) 24

26. A licensee seeking a waiver of the CE requirement shall apply to the Board in writing a least ____ days prior to renewal?
    A) 30   B) 60   C) 90   D) 180

27. Every license and business permit holder shall notify the Board within ____ days of a change of address of record.
    A) 7   B) 10   C) 25   D) 30

28. Every license and business permit holder shall notify the Board ____of a change in name of the electrical contracting business.
    A) within 7 days   B) within 10 days   C) within 25 days   D) immediately

Subchapter 2: License & Business Permit Requirements

29. How long must an applicant wait before he/she is eligible to re-take failed sections of the licensing examination?
    A) 3 months   B) 6 months   C) 1 yr.   D) 6 weeks

30. If an applicant fails to obtain a passing score of their second attempt the applicant shall wait ____ prior to retaking the failed section of the exam.
    A) 3 months   B) 6 months   C) 1 yr.   D) 6 weeks

31. If an applicant fails to obtain a passing score on their third attempt they shall_____
    A) be required to retake ALL exams over.
    B) be required to wait 1 yr prior to retake the failed sections.
    C) be required to wait 6 months prior to retaking the failed section.
    D) be required to re-apply to the Board for approval to retake exam.

32. Every licensee shall renew their license within ____ of expiration or have their license suspended.
    A) 20 days   B) 10 days   C) 90 days   D) 30 days

Subchapter 3: Standards of Practice

33. Every licensee holding a business permit shall obtain _____.
    A) Workers Compensation
    B) Surety Bond
    C) Liability Insurance
    D) All the above

34. For a business who's qualified licensee is rendered incapable of fulfilling their professional duties the may grant a delay in returning the licensee's for a maximum of _____
    A) 30 days   B) 90 days   C) 180 days   D) 360 days

35. As an electrical contractor you are required to provide constant on-site supervision for electricians employed with less than _____ experience working under the UCC.
    A) 3 ½ yrs   B) 2 ½ yrs   C) 1 ½ yrs   D) 1 yr

36. As an electrical contractor you are required to provide on-site supervision for electricians employed and can employ a qualified journeyman to perform this task provided _____.
    A) the journeyman is on-site providing constant supervision for electricians with less than 3 ½ yrs experience
    B) the journeyman provides directions and inspects work of electricians with more than 3 ½ yrs experience
    C) A & B
    D) they have a license (Journeyman cannot be delegated supervisory requirement)

37. A qualified journeyman electrician can supervise up to ____ electricians with 6 months or less experience.
    A) 2   B) 5   C) 10   D) not allowed

38. When two or more persons form a joint venture for the purpose of contracting to perform electrical work, then ....
    A) each party must be licensed
    B) each party must hold an electrical business permit
    C) at least one person must be licensed
    D) at least one person must hold a electrical business permit

39. Which of the following is NOT governed by the rules and regulations of the State Board Of Examiners.
    A) Your pricing B) Your continuing education C) Workers supervision D) Marketing practices

40. Which of the following does **NOT** met the minimum requirements for a business to engage in electrical installation of telecommunication equipment?
    A) An electrical contractor's license
    B) An exemption from the N. J. Board of Examiners
    C) An electrical business permit
    D) All of the above met the minimum requirements

41. Business's holding only a telecommunication exemption shall not perform which of the following?
    A) Installation of telecommunications wiring within a premise
    B) Installation of 120V power between power supply integral with the telecommunications equipment
    C) Installation of telecommunications in hazardous locations
    D) Installation of telecommunications wiring between computer equipment

42. For a licensed well driller / installer which of the following electrical task can he **NOT** perform?
    A) Installation of 240V, 1ϕ exterior underground wiring from the controller to the well pump
    B) Installation of 6ft of 120V, 1ϕ interior wiring from the OCP device to the pump disconnect
    C) Installation of 8ft of 12-2 Romex run inside to connect the pump to the pressure switch
    D) Installation of 240V, 1ϕ exterior branch circuit wiring to an exterior controller

43. Which of the following electrical tasks are exempt from the license and business permit requirements of the Board of Examiners of Electrical Contractors?
   A) Replacing a component integral to electrical equipment
   B) Replacement of a lighting ballast
   C) Replacement of an LED driver
   D) All the above

44. A "Qualified Journeyman" has acquired _____ of practical experience and ___ classroom hours of related instruction.
   A) 6 months / 2 years   B) 4000 hrs / 544 hrs   C) 5 years / 2 yrs   D) 8000 hrs / 576 hrs

45. A "Qualified Journeyman" has to complete _____ hours of continuing education to renew his registration.
   A) 34   B) 24   C) 10   D) none required

Appendix H: Fire Alarm, Burglar Alarm and Locksmith Licensees and Business

46. For a Fire or Burglar Alarm applicant which of the following is **NOT** a requirement?
   A) Have completed 6720 hours of hands-on work in the field
   B) Having not been convicted of the 4th degree offense of engaging in the unlicensed practice of electrical contracting
   C) Have completed 80 hours of specified training courses.
   D) None of the above

47. A licensed Fire Alarm contractor may have an employee perform unsupervised work provided the employee _____ ?
   A) Has three years in-field practical experience
   B) Has completed the Board approved training course requirements of 13:31A-3.6
   C) A & B
   D) Unsupervised work is not allowed as per 13:31A-3.7

48. For every___ employees who are working on Burglar and Fire alarm rough wiring installation requires "Supervision".
   A) 5   B) 10   C) 12   D) 20

Appendix I: Professional and Occupation Uniform Enforcement Act

49. Any person violating the Professional and Occupation Uniform Enforcement act shall be guilty of....
   A) a 4th degree felony.
   B) a crime with the severity based on the Boards findings.
   C) a misdemeanor.
   D) Misconduct.

50. Any person violating the Professional and Occupation Uniform Enforcement act shall be liable to a civil penalty of not more than _____ for the first violation.
   A) $5000
   B) $7500
   C) $10000
   D) $20000

Chapter 4 Project Risk Management

Main Topics for This Section
   A. Fundamentals of insurance
      - Workers' Compensation Insurance
      - Unemployment Insurance
   B. Fundamentals of bonds

The chapter reviews Chapter 4: Managing Risk of the NASCLA Contractors Guide to Business, Law, and Project Management New Jersey Edition.

Supplementary Notes

This section deals primarily with how insurance and bonds are used to manage risk. One important risk not addressed is financial risk with respect to payment schedules and cash flow. As a contractor you must evaluate the financial risk of the job. That is the risk taken in expending time and money with respect to the payment schedule. Doing all the work and putting up all the money up-front prior to any payment is a substantial risk. All contracts should be negotiated to have mutual risk. This means your labor and the owner's money to cover material or a progressive payment schedule that you are financially comfortable with. Many general contractors will expect you to complete most of the work with your labor and money and for the lowest bid. If the risk is high so should your profit margin. However if the risk is low a lower profit margin is reasonable. An example would be to bid a job with a tight profit margin for a customer if payment is guaranteed or bid the job with a higher profit margin if the payment schedule is end-loaded. Don't get mesmerized with the potential profits or promises for more work and allow yourself to be put financially at risk now.

A point on insurance coverage as a new electrical contractor; do not forget the tool rider on your business insurance policy. As a contractor your truck can be carrying around $10,000 to $30,000 dollars' worth of tools most of the time. This is a big risk if you are not insured. Most tool riders can provide coverage for a major lost at a reasonable fee.

Supplemental definitions:

Insurance rider: This is the term referring to special coverage for a specific risk when added to an insurance policy.

Exclusions: The section of an insurance policy that states various conditions for which the insurance policy will not provide coverage.

Attractive Nuisance: The term describing a job sites attractive and inviting nature invoking special liability obligations on the contractor to protect children trespassing on the construction site.

Employer's Liability insurance provides coverage over and above Workers' Compensation in case of judgments stemming from civil actions due to injury or death of an employee.

Surety Bonds: Bonds are similar to insurance in that they provide financial reparation for specific damages. However surety bonds differ from insurance in that the bond does not protect the contractor. Instead, the contractor remains responsible for the financial liability incurred by the surety bond company in the event of a claim. In the event of a claim, statutory bonds need not specifically cite the claim but have it covered under the statue. Whereas for common-law bonds the claim must be specifically called out in the written bond.

Fundamentals of Insurance

1. _____ is a protective measure that reduces the contractor's losses due to exposure to risk.
   A) Surety Bond   B) Insurance   C) Safety program   D) Miller Act

2. When special coverage's are attached to an insurance policy these attachments are called _____.
   A) riders   B) exclusions   C) addenda's   D) floaters

3. A _____ insurance policy protects the contractor against any physical loss or damage to the project or project materials, except for listed exclusions.
   A) Liability   B) All-Risk Builder's Risk   C) Named Peril Builder's Risk   D) Equipment Floater

4. The section of an insurance policy that states various conditions for which the insurance policy will not provide coverage is known as the _____
   A) liability   B) exclusions   C) indemnification   D) floater

5. A _____ insurance policy protects the contractor against losses to the project due to fire and lightning only.
   A) Liability   B) All-Risk Builder's Risk   C) Named Peril Builder's Risk   D) Equipment Floater

6. _____ on the property insurance policy provides coverage against damage to property which belongs to the contractor while being transported.
   A) Liability   B) Automobile   C) Transportation Floater   D) Equipment Floater

7. _____ is the term describing a job sites attractive and inviting nature invoking special liability obligations on the contractor to protect children trespassing on the construction site.
   A) Indemnification   B) Attractive Nuisance   C) Invitational Liability   D) Attractive Premise

8. _____ insurance provides protection when a contractor is liable for the acts of others for whom he has responsibility.
   A) Contractual Liability
   B) Completed Operations Liability
   C) Contractor's Protective Public and Property Damage Liability
   D) Professional Liability

9. _____ insurance provides protection to a contractor from liability arising out of errors or negligent acts in performing design duties.
   A) Contractual Liability
   B) Completed Operations Liability
   C) Contractor's Protective Public and Property Damage Liability
   D) Professional Liability

10. _____ insurance provides protection to the contractor from liability to third parties arising after the project has been completed.
    A) Contractual Liability
    B) Completed Operations Liability
    C) Contractor's Protective Public and Property Damage Liability
    D) Professional Liability

11. _____ insurance provides protection when a contractor is liable for their negligence or that of the owner, architect, or other parties by the terms of the contract while under written contract.
A) Contractual Liability Insurance
B) Completed Operations Liability
C) Contractor's Protective Public and Property Damage Liability
D) Professional Liability

12. _____ insurance provides coverage over and above Workers' Compensation in case of judgments stemming from civil actions due to injury or death of an employee.
A) Contractual Liability Insurance
B) Employee Operations Liability
C) Contractor's Protective Public and Property Damage Liability
D) Employer's Liability

13. A _____ is the right granted by the contractor to the insurance company to cover negligent damages caused by other parties.
A) liability   B) subrogation   C) indemnification   D) blanket coverage

14. When the owner of a small or mid-size Construction Company chooses to bundle property and liability coverage together under one policy is know as _____.
A) Contractual Liability Insurance
B) Complete Operations Liability
C) Business Owner's Policy
D) Wrap-Up Liability

15. What type of insurance policy is designed to increase the dollar and coverage limits of underlying insurance?
A) Contractual Liability
B) Umbrella
C) Contractor's Contingent
D) Employer's Liability

16. What type of mandatory insurance is managed by the federal government?
A) Social Security
B) Workers Compensation
C) Employee's liability
D) FUTA

17. What type of insurance is mandatory for a contractor who employs employees and benefits are accrued through non-voluntary employee and employer contributions?
A) Social Security
B) Workers Compensation
C) Employee's liability
D) FUTA

18. What type of mandatory insurance has its premiums fully paid by the employer based on occupational risk and company business dollar volume?
A) Social Security
B) Workers Compensation
C) NJ State Unemployment
D) FUTA

19. Which of the following is **NOT** covered by social security insurance?
    A) retirement benefits   B) survivor benefits   C) long term disability benefits   D) medical benefits

20. Which of the following kinds of insurance are **NOT** legally required by electrical contractors?
    A) Health   B) Unemployment   C) Liability   D) Workers' Compensation

Fundamentals of Bonds

21. The primary difference between Bonds and Insurance with respect to liability is?
    A) Surety Company assumes all liability for a bond
    B) Insurance Company assumes no liability
    C) The contractor retains liability with a bond
    D) A&B

22. Private projects are covered by _____ bonds?
    A) Federal   B) State   C) statutory   D) common-law

23. A claim under a _____ bond need not be specifically outlined in the bond.
    A) private
    B) common-law
    C) statutory
    D) N/A all bonds address only provisions written into the bond

24. _____ is a bond protecting the owner from loses associated from a successful bidder not honoring the bid.
    A) Bid Bond   B) Holding Bond   C) Performance Bond   D) Lien Bond

25. _____ is a bond guaranteeing the owner that the project will be completed according to the specifications of the contract.
    A) Bid Bond   B) Payment Bond   C) Performance Bond   D) Lien Bond

26. What type of bond does not protect the contractor and results in the contractor being liable for losses sustained from the Bonding Company.
    A) Surety Bond   B) Performance & Payment Bonds   C) Lien Bonds D) All Bonds

27. _____ is a bond that protects the owner, developer, and contractor against liens filed by unpaid parties to the work.
    A) Bid Bond   B) Payment Bond   C) Performance Bond   D) Lien Bond

28. _____ is a bond that provides coverage against defects in workmanship or materials for a stated period.
    A) Maintenance Bond   B) Warranty Bond   C) Performance Bond   D) Lien Bond

29. The Bonding Company when it underwrites a bond agrees to cover liabilities for conditions _____.
    A) Specified in the original contract only
    B) Project changes agreed to in writing by the owner and contractor
    C) A & B
    D) For any and all project conditions and modifications agreed to by the owner

30. The Miller Act applies to which type of contracts?
     A) Federal only   B) Federal and State   C) All   D) Private and Commercial

31. The Miller Act applies to contract over _____ amount.
     A) $10,000   B) $100,000   C) $250,000   D) $500,000

32. The Miller Act requires a contractor to secure _____.
     A) Surety Bond   B) Performance & Payment Bonds   C) Liability Insurance D) Lien Bond

33. The Miller Act requires a payment bond in the sum of _____ for a $2M project.
     A) equal to Performance Bond
     B) not to be greater than $2.5 million dollars
     C) $800,000 dollars
     D) $2,000,000 dollars

34. States statues that impose similar surety laws as the Miller Act for public work projects are referred
     to as _____.
     A) Federal Surety laws
     B) Common-Law surety statues
     C) "Little Miller Acts"
     D) Construction Surety Act

35. The Construction Industry Payment Protection Act of 1999 requires a payment bond in the sum of
     _____ for a $2M project.
     A) equal to Performance Bond
     B) A&D
     C) $800,000 dollars
     D) $2,000,000 dollars

Chapter 5 Estimating & Bidding

Main Topics for This Section
    A. Bidding Process & Documents
    B. Project Documents
    C. Estimating Methods

The chapter reviews Chapter 7: Bidding and Estimating of the NASCLA Contractors Guide to Business, Law, and Project Management New Jersey Edition.

Supplementary Notes

    Estimating and bidding is a combination of a special skill and business savvy. A good understanding of the code and current industry materials allows a good estimator to determine the lowest cost to complete a project and meet all engineering requirements and the NEC. That's the skill. Then the business savvy is needed in turning a good estimate into a profitable bid though an understanding of the customer and the project. The reference manual provides some basic insight to the fundamentals of estimating and bidding but these skills are only truly achieved though practical experience.

Additional Definitions:

General Conditions: The section of the project documents that outline the roles of the owner, architect, and engineer.

Project Drawings: This section of the project documents shows the physical aspects of the project including dimensions, quantity, size, and location of items.

Direct Labor Cost: The labor cost for efforts specifically and directly associated with the project.

Indirect Labor Cost: Also known as Labor Burden cost are all other labor cost that are not associated with direct efforts on the project. These would include employer taxes, workers' compensation cost, insurance cost, company benefits, and training cost.

Bid Process & Documents

1. Which bid document gives a brief overview of the project, deadlines and general requirements?
   A) Invitation to Bid   B) Bid Instructions   C) General Conditions   D) Supplements

2. Which bid document defines how the bid is to be completed and submitted?
   A) Invitation to Bid   B) Bid Instructions   C) General Conditions   D) Supplements

3. Which bid document would the bidder find information on property survey and soil analysis?
   A) Invitation to Bid   B) Bid Instructions   C) General Conditions   D) Supplements

4. If changes are issued to a bid package prior to the bid submittal date a(n) _____ is issued officially and becomes part of the bid package?
   A) Supplement   B) bid form   C) addendum   D) contract modification

5. Which section of the project documents would a bidder find information on designated specified dollar amounts for specific items to be included in the bid?
   A) Invitation to Bid   B) Allowances   C) General Conditions   D) Supplements

6. When a general contractor utilizes and/or reveals one subcontractors bid to another subcontractor in an attempt to obtain a lower price, this practice is known as _____.
   A) Bid Shopping   B) Bid Peddling   C) Bid Rigging   D) Low Bid Marketing

7. _____ is when a subcontractor approaches the general contractor post bid award with a revised lower bid in an attempt to capture the bid award.
   A) Bid Shopping   B) Bid Peddling   C) Bid Rigging   D) Low Bid Marketing

8. When contractors conspire together to win, share, or price fixing a bid this is known as _____.
   A) conspiracy   B) price rigging   C) fixed bid   D) bid rigging

Project Documents

9. What project document section defines payment limitations for specified project items such as lighting fixtures?
   A) General Conditions   B) Supplements   C) Allowances   D) Specifications

10. What project document section defines changes to the released bid package issued prior to bid?
    A) Supplementary Conditions B) Supplements C) Allowances   D) Addenda

11. What project document section outlines the roles of the owner, contractor, architect, and engineer?
    A) Supplementary Conditions B) General Conditions C) Allowances   D) Addenda

12. What project document section defines details of construction methods, types of materials and workmanship requirement?
    A) Supplementary Conditions B) Supplements C) Specifications   D) Addenda

13. What project document section shows physical aspects of the project?
    A) Drawings   B) Supplements   C) Specifications   D) Addenda

Methods of Estimating

14. A quantity survey listing of all materials and items of work required for a project is also known as a
_____?
A) Takeoff   B) Unit Price Method   C) Approximation Method   D) itemization

15. What is the method of estimating called when the estimate is based on a listing all materials and labor needed for the project and assigning a specific price to each?
A) Quantity Take-off Method B) Unit Price Method C) Approximation Method D) SWAG Method

16. Conduit delivery costs are included under what part of the estimate?
A) Material Cost   B) Labor Cost   C) Equipment Cost   D) Subcontractors

17. What cost is health insurance charged to in an estimate?
A) Company Overhead   B) Direct Labor Cost   C) Indirect Labor Cost   D) Project Overhead

18. What cost is equipment finance cost charged to in an estimate?
A) Company Overhead   B) Material Cost   C) Equipment Cost   D) Project Overhead

19. What cost is Worker's Compensation insurance charged to in an estimate?
A) Company Overhead   B) Direct Labor Cost   C) Labor Burden Cost   D) Project Overhead

20. Legal fees related to the bid and contracts are charged as?
A) Company Overhead   B) Direct Labor Cost   C) Indirect Labor Cost   D) Project Overhead

21. What is the cost, added to a bid to cover a contractor's uncertainty in factors influencing project completion?
A) Profit   B) Contingency   C) Overhead   D) SWAG

22. If your estimate included $3000 for direct cost and $2500 for indirect cost with a Profit margin of 10% of total project cost, what would be the total bid price submitted?
A) $550   B) $5500   C) $6050   D) $5800

23. What is the method of estimating called when the estimate is based assigning costs to each type of unit of construction?
A) Detailed Survey Method B) Unit Price Method C) Approximation Method D) SWAG Method

24. What is the method of estimating called when the estimate prepared by the architect using cost models?
A) Detailed Survey Method
B) Unit Price Method
C) Approximation Method
D) Conceptual Method

Chapter 6 Contracts

Main Topics for This Section
      A. Elements of a Contract
      B. Responsibilities of the contractor and owner
      C. Payment methods
      D. Types of contracts

The chapter reviews Chapter 8: Contract Management of the NASCLA Contractors Guide to Business, Law, and Project Management New Jersey Edition.

Supplementary Notes

Primary rule for business is " get it in writing prior to starting the work ". Do not hesitate to discuss payment schedules with customers or general contractors. All experienced contractors have fallen into the trap where a customer is vague or is in a hurry to complete the job. In your haste to mee the customer's needs, little or nothing is written down and signed between the parties involved. In mos cases payment for your services is never received. The best advice is to write down in simple language a description of the work, expected completion date and payment schedule as a minimum. In addition the payment amount(s) must be written both as a number an in text. The text amount will take precedence over the number. For the work description, a simple diagram of the installation with an outline of all major components should be drawn up and signed by all parties along with a written description of the components and circuits. This will provide a basis to settle disputes between the contractor and the owner or general contractor.

Additional Definitions:

Time is of the essence: means all performance must be completed in a timely manner even if the completion date is not specified in the contract.

Elements of a Valid Contract

1. Which of the following is required n order for a contract to be enforceable?.
    A) "Consideration" B) Completion Dates C) Specific Description of the work D) Owner's name

2. Which of the following is NOT a valid and enforceable contract?
    A) Written contract only
    B) Oral contract only
    C) Addendum's to a written contract
    D) Written contract agreeing to perform electrical work not to NEC standards

3. The element of a contract that specifically outlines the obligations of the contract including the work and compensation is known as the _____.
    A) Acceptance
    B) Consideration
    C) Offer
    D) Obligation Terms

4. If an offer does not have specific terms for expiration then the offer is said to expire within _____.
    A) 7 working days   B) 30 days from submittal   C) "reasonable time"   D) never

5. Which of the following constitutes a contractor acceptance of a contract?
    A) Signed offer
    B) Proposed offer
    C) Counter offer
    D) All the above

Contract Provisions

6. _____ are partial payments made after specific phases of the construction are complete?
    A) Retainage Payment
    B) Progress Payment
    C) Final Payment
    D) Cost Plus Payment

7. The withholding of a percentage of payment from the contractor until final completion and acceptance of the work is known as _____?
    A) a warranty   B) retainage   C) indemnification   D) limited liability

8. What part of a contract specifies the rights, duties and the responsibilities of the contractor?
    A) General Conditions   B) Boilerplate Provisions   C) Indemnification   D) Addenda

9. Which provision of the contract modifies the general conditions defining project specific requirements?
    A) General Provisions B) Indemnification   C) Warranties   D) Supplemental Conditions

10. Which provision of the contract contains the standard language or clauses used in a legal contract?
    A) General Provisions B) Indemnification   C) Miscellaneous Clauses   D) Force Majeure

11. When one of the parties involved with a contract fails to perform in accordance with the terms and conditions of the provisions it is defined as _____.
   B) Civil Contract Dispute   B) Indemnification   C) Breach of Contract   D) Force Majeure

12. A _____ is a "Breach of Contract" which results in a serious violation that voids or terminates the contract.
   A) Immaterial Breach   B) Material Breach   C) Partial Breach   D) Contractual Liquidation

13. A _____ is a "Breach of Contract" which may result in a termination of the contract or the injured party suing for only the value of the damages incurred.
   A) Immaterial Breach   B) Material Breach   C) Indirect Breach   D) Contractual Liquidation

14. When a contract imposes monetary damages for failure to meet specific provisions of the contract, these damages are known as _____.
   A) Performance Penalties
   B) Liquidated Damages
   C) Breach of Contract
   D) Performance Forfeitures

15. The provision of a contract which absolves the owner or contractor from damages related to "acts o god" is known as _____.
   A) Indemnification   B) Liquidated Damages   C) Warranty   D) Force Majeure

16. What is the provision called in a contract that stipulates certain parties are not liable for damages incurred by third parties?
   A) Force Majeure   B) Retainage   C) Allowances   D) Indemnification

17. What is the provision called in a contract that defines the rights of the owner to require the contractor to correct faulty work or material?
   A) Warranties
   B) Force Majeure
   C) Certification Of Completion
   D) Indemnification

18. _____ means all performance must be completed in a timely manner even if the completion date is not specified in the contract.
   A) Time is of the essence   B) Substantial completion
   C) Notice requirements   D) Progress schedule

19. What is the provision called in a contract that contains language that provides background information to the contract?
   A) Warranties
   B) Force Majeure
   C) Recitals
   D) Indemnification

20. Which of the following contract provision may not be legally enforceable?
   A) Warranties
   B) Force Majeure
   C) Recitals
   D) Indemnification

Types of Construction Contracts & Methods

21. What type of contract is one in which the contract is based on a fixed fee for the contractor and the owner agrees to pay for all other construction cost with the total cost unknown until completion of the project?
    A) Lump-Sum   B) Unit-Price   C) Cost Plus Fee   D) Turn Key

22. What type of contract is one in which the contract is based on the cost of a unit of the project and the total number of units to be completed may not be stated?
    A) Lump-Sum   B) Unit-Price   C) Cost Plus Fee   D) Turn Key

23. What type of contract is one in which the contract stipulates that the contractor is obliged to completed the project for a fixed price?
    A) Lump-Sum   B) Unit-Price   C) Cost Plus Fee   D) Turn Key

24. What type of contract relationship is defined as a single company being responsible for the entire design and construction process but not financing or land acquisition?
    A) Design to Cost   B) Fast Track Design   C) Design / Build   D) Turn Key

25. What type of contract relationship has the contractor being responsible for the entire design, construction process and all financial and land acquisition responsibilities?
    A) Design to Cost   B) Fast Track Design   C) Design / Build   D) Turn Key

26. What type of contract relationship has the contractor beginning the construction prior to the completion of contract documents?
    A) Design to Cost   B) Fast Track Design   C) Design / Build   D) Turn Key

27. Change orders to a contract prior to award are called _____.
    A) Addenda   B) Contract Addition   C) Modifications   D) Meditation

28. Change orders to a contract after contract award are called _____.
    A) Addenda   B) Contract Addition   C) Modifications   D) Meditation

29. The processes of settling contract disputes without costly judicial trails are known as _____.
    A) Litigation
    B) Contract Litigation Resolution
    C) Alternative Contract litigation
    D) Alternative Dispute Resolution

30. Which of the following is the 1st step in the ADR process where unstructured discussions are used for dispute resolution?
    A) Mediation   B) Negotiation   C) Collaborative Law   D) Arbitration

31. Which of the following ADR processes utilize an independent third party who is trained to facilitate and agreement between contract parties to resolve the dispute?
    A) Mediation   B) Negotiation   C) Collaborative Law   D) Arbitration

32. Which of the following ADR processes is a facilitative process where parties agree to resolve the dispute without litigation thru discussions between their advocates?
A) Mediation   B) Negotiation   C) Collaborative Law   D) Arbitration

33. Which of the following ADR processes uses a third party to act as a judge and render a legally binding decision to resolve the dispute?
A) Mediation   B) Negotiation   C) Collaborative Law   D) Arbitration

Chapter 7 Project Management

Main Topics for This Section
- A. Fundamentals of project planning and scheduling
- B. Management supervision functions
- C. Project progress monitoring concepts and methods

The chapter reviews Chapter 9: Scheduling and Project Management of the NASCLA Contractors Guide to Business, Law, and Project Management New Jersey Edition.

Supplementary Notes

None

Additional Definitions:

Project Planning: This is the process of defining and analyzing each of the tasks associated with the project

Gnatt Chart: An addition tern used for the critical path method bar chart.

Production Rate: Defines the rate at which a task is performed. This is a method to monitor progress and cost against the estimate. It is defined as task per unit of time.

Scheduling Process

1. In managing a project the process of defining and analyzing each of the required project tasks is called _____.
   A) CPM   B) project scheduling   C) labor management   D) project planning

2. _____ is the amount of time added to the estimated task duration to plan for unexpected delays.
   A) Earliest Time   B) Float Time   C) Contingency Time   D) Padding

3. Calculating the _____ ranges for each task provides a guide to determine how on schedule a task or the entire project may be.
   A) activity duration   B) labor charge   C) earliest & latest time   D) float time

4. _____ is the amount of time left over based on the estimated task duration time, after a task is completed.
   A) Earliest Time   B) Float Time   C) Contingency Time   D) Padding

5. The amount of time an activity can be delayed without impacting the earliest start time of the next task is called _____.
   A) open time   B) free float   C) contingency time   D) padding

6. The activities with no or zero float time are considered _____ activities.
   A) critical   B) free float   C) effective scheduled   D) key

Scheduling Methods

7. Which of the following scheduling methods indicate the interdependencies of the activities.
   A) Calendar Scheduling
   B) Bar Chart
   C) Critical Path Method
   D) All the above

8. What type of planning and scheduling analysis method produces a diagram which depicts the longest continuous sequence of activities in a project?
   A) CPM   B) Gantt   C) Pie   D) Line

9. The _____ is the sequence of tasks that determines the duration of the project.
   A) earliest duration   B) Gantt   C) float time   D) critical path

10. _____ refers to the amount of cash available after liabilities or debts are paid.
   A) Working Capital   B) Cash flow   C) Earn Value   D) Cash management

11. _____ outlines the stages of construction with anticipated revenues and expenditures.
   A) Working Capital schedule
   B) Cash flow budget
   C) Earn Value
   D) Cash management

Project Management

12. _____ is the individual responsible for the management of a specific area or trade of the project.
A) Forman   B) Project Manager   C) Architect   D) Expediter

13. _____ is the individual responsible for the management for all phases of work on the project.
A) Forman   B) Project Manager   C) Architect   D) Expediter

14. _____ is the individual responsible for the management of daily onsite operations of the project.
B) Forman   B) Project Manager   C) Superintendent   D) Expediter

15. _____ is the individual responsible for timely delivery of all materials work on the project.
C) Forman   B) Project Manager   C) Architect   D) Expediter

Cost and Quality Control

16. A cost control process that allows for material delivery to arrive when needed for use and not before is called_____.
A) Purchase Order   B) Just-In-Time   C) Receiving   D) Expediter

17. As part of a quality control process the _____ should review and approve shop drawings.
A) Foreman   B) Architect   C) Project Manager   D) Expediter

18. A contractor may develop _____ to detail specific aspects of a project.
A) Purchase Order   B) Contract Specifications   C) Shop Drawings   D) Blue Prints

19. To clarify an architect's plans and specifications _____ maybe required to be produced by the material supplier or contractor.
A) Work Specs   B) Cut Sheet   C) Contracts Letter of Clarification   D) shop drawings

20. _____ is a project management approach where the owner provides incentives to the contractor for cutting cost without sacrificing quality.
A) Quality Assurance   B) Value Engineering   C) Cost Control   D) Earn Value

21. What is the production rate (# outlets installed/hour) for a job that had 3 electricians rough-in a residential unit (service not included), which had 88 electrical outlets, over 3 (8-hour) day period?
A) 3.7   B) 29.2   C) 1.22   D) 2.33

Chapter 8 Employee Management

Main Topics for this Section
    A. Employee hiring laws
    B. Employment laws
    C. Employment documentation
    D. Employee required coverages

The chapter reviews Chapter 11: Employee Management of the NASCLA Contractors Guide to Business, Law, and Project Management New Jersey Edition.

Supplementary Notes

    An important point not emphasized in this chapter but will most likely show up on the state test is the computation of gross salary with straight and over-time pay included. The following outlines the method and rules for computing over-time pay.
- For nonexempt employees overtime pay shall be computed at a rate of 1 ½ times the regular hourly rate
- Overtime shall be paid for all hours worked over the first 40 hours of work in a workweek.
- Even if employees are paid bi-weekly, overtime is computed based on the hours worked in each week.

Example 1 :   What is the gross salary for an employee who is paid $20 an hour and worked 45 hours last week.

    $20 /hr for first 40 hours   = $800
    $20 /hr *1 ½ * 5 hours overtime = $150
         Total Gross Salary = $950

Example 2 :   What is the gross bi-weekly salary for an employee who is paid $20 an hour and worked 45 hours the first week and 30 hours the second week.

    $20 /hr for the first 40 hours (week #1)   = $800
    $20 /hr *1 ½ * 5 hours overtime (week #1)   = $150
    $20 /hr for the first 40 hours (week #2): $20 * 30 hr = $600

         Total Gross Salary = $1550

Americans with Disabilities Act (ADA)

The ADA allows for an employer to ask an applicant for a job about their ability to perform the job tasks require a medical exam after making a job offer, and require an applicant to take a drug test prior to a job offer.

Employee Hiring

1. Which of the following interview questions maybe use during an interview and would not result in a legal action against you or your company?
   A) What ethnic background are you?
   B) Are you married?
   C) What was your reasons for leaving the last job?
   D) None of the above

2. Which of the following interview questions is **NOT** recommended for use during an interview and may result in a legal action against you or your company?
   A) Do you have any disabilities hindering you meeting the job requirements?
   B) What is your ability and education level to work electrical construction?
   C) What was your reasons for leaving the last job?
   D) All of the above

3. What of the following federal law requires reporting of new hires to the State?
   A) The Federal Reporting Work Act of 1996
   B) The NJ New Hire Reporting Program
   C) The Work and Opportunity Reporting Act of 1996
   D) The Personal Responsibility and Work Opportunity Reconciliation Act

4. NJ employers are required to report all new hires no later than _____ days from date of hire/re-hire to the NJ New Hire Reporting Program.
   A) 5 days   B) 7 days   C) 10 days   D) 30 days

5. NJ employers who submit hiring reports magnetically or electronically, to the NJ New Hire Reporting Program, must submit the transmissions ____monthly and no longer than _____ days apart.
   A) Twice - 5   B) Four times - 7   C) Twice - 16   D) Twice - 30

6. NJ employers who fail to submit new hire information as required to the NJ New Hire Reporting Program shall be subject to a fine up to ____ per incident.
   A) $25   B) $500   C) $300   D) $2500

7. A hired employee under the age of _____ must obtain a work permit from the State of NJ.
   A) 18   B) 21   C) 16   D) 14

8. The NJ Child Labor prohibits work of child under the age of _____ except in very limited circumstances.
   A) 14   B) 18   C) 21   D) 16

Employee Documentation

9. What form is required to be completed by the employee to show they have legal immigration status to work in the Untied States?
   A) W-4   B) W-5   C) NJ W-4   D) I-9

10. What form is required to be completed by the employee to define the appropriate level of federal tax withheld?
    A) W-4   B) W-5   C) NJ W-4   D) I-9

11. What form is required to be completed by the employee to received earned income credit in advance?

   B) W-4 B) W-5 C) NJ W-4 D) I-9

Key Labor Laws

12. What is the federal minimum wage effective July 24, 2009 for non-exempt employees?

   B) $8.25 B) $7.10 C) $7.25 D) $9.75

13. How much gross salary is to be paid to an hourly worker who earns $10/hr and has worked 50 hours in this work week?

   A) $500 B) $750 C) $550 D) $400

14. Employees under 20 years of age can be paid ____ per hour for the first 90 consecutive days as the federal minimum wage?

   A) $5.05 B) $5.15 C) $4.25 D) $4.75

15. How much bi-weekly gross salary is to be paid to an hourly worker who earns $10/hr and has worked 50 hours in the first week and 20 hours in the second week?

   A) $850 B) $550 C) $750 D) $700

16. Which of the following employment practices are **NOT** regulated by the Fair Labor Standards Act?

   A) Premium pay for weekend
   B) Basic minimum wage
   C) Overtime pay standards
   D) Employment of minors

17. The Fair Labor Standards Act requires employers to keep employment records which must include which of the following employment data?

   A) Basis on which employee wages are paid
   B) Total hours worked each workday
   C) Total wage paid each pay period
   D) All the above

18. Willful violations of the employment laws under FLSA may result in fines up to _____.

   A) $1,100 B) $10,000 C) $1000 D) $100,000

19. Employers must keep an I-9 form on file for at least ____ from the date of employment.

   A) 6 months B) 1 year C) 3 years D) 5 years

20. Employers must complete an I-9 form with required documentation within ____ days from the date of employment.

   A) 3 B) 7 C) 10 D) 22

21. Which document below **DOES NOT** establish employment eligibility within the U.S. ?

   A) U. S. Social Security Card
   B) Original / Certified Copy of birth certificate
   C) Drivers License with Photo I.D.
   D) U.S. Citizen ID card

22. Which document below establishes identity and employment eligibility within the U. S. ?
    A) U. S. Passport
    B) Certification of U.S. Citizenship
    C) Certification of Naturalization
    D) All the above

Americans with Disabilities Act (ADA)

23. It is unlawful to _____
    A) to ask an applicant for a job about their ability to perform the job tasks
    B) to require a medical exam after making a job offer
    C) require an applicant to take a drug test prior to a job offer
    D) require an applicant to have a medical exam prior to making a job offer

Other Key Labor Laws

24. What act requires payment of minimum wage rates and overtime pay on federal financed construction contracts?
    A) Davis-Bacon   B) Walsh-Healey Public Contracts   C) Service Contract   D) None

25. What act requires payment of minimum wage rates and overtime pay on contracts to provide services to the federal government?
    A) Davis-Bacon   B) Walsh-Healey Public Contracts   C) Service Contract   D) None

26. What act requires payment of prevailing wage rates and fringe benefits on contracts to provide services to the federal government?
    A) Davis-Bacon   B) Walsh-Healey Public Contracts   C) Service Contract   D) None

27. What law prohibits an employer from firing an employee whose pay is garnished for payment of a single debt?
    A) Wage Garnishment
    B) Employee Protection
    C) Employee Garnishment
    D) Fair labor Act

28. Which of the following Labor Laws are NOT applicable within New Jersey State?
    A) Wage Garnishment Law
    B) Uniformed Services Employment and Reemployment Rights Act
    C) Worker Adjustment and Retraining Notification Act
    D) Right-To-Work Laws

29. The Family and Medical Leave Act entitles eligible employees to take up to _____ weeks unpaid job-protected leave each year, with maintenance of group insurance for specific family and medical reasons.
    A) 12   B) 2   C) 9   D) 4

30. Which of the following labor posters are NOT required to be posted by NJ State law?
   A) Gender Equality Notice
   B) Safety and Health Protection on the Job
   C) NJ Wage and Hour Law
   D) NJ SAFE Act

Employment Benefits

31. The cost of Workers' Compensation is paid for through _____.
   A) deductions   B) employee payments   C) employer payments D) N. J. State taxes

32. The failure to provide workers' compensation insurance if required is considered a _____ if determined to be intentional.
   A) Fourth degree crime
   B) First degree felony
   C) Misdemeanor offense
   D) Disorderly Persons offense

33. The cost of Unemployment Compensation is paid for through _____.
   A) deductions
   B) employer payroll tax
   C) Worker's Compensation Insurance
   D) Division of Worker's Compensation

34. Unemployment Compensation insurance tax must be paid by the employer if they had paid wages in excess of _____ within any quarter of the year.
   A) $1000   B) $1500   C) $2500   D) $5000

35. Which of the following determines the Unemployment Compensation insurance tax rate for an employer?
   A) Based on wages paid on a calendar quarter basis
   B) experience rating
   C) unemployment benefits claims
   D) all the above

36. The Unemployment Compensation tax rate for a newly acquired business is based on _____.
   A) the new owners rating
   B) the previous owners rating
   C) the new employer rating
   D) a re-computed combination of the new owner and previous business rating

37. _____ is the practice of transferring of employees between businesses for the purpose of obtaining lower unemployment compensation tax rate.
   A) Deductions   B) Withholding   C) COBRA   D) SUTA dumping

38. The _____ provides the right to continue health care coverage for employees at their own expens if due to a qualifying event that terminates employer health insurance.
   A) COBRA   B) SUTA   C) HIPAA   D) FICA

39. As an employer you must be aware of the _____ act that addresses an employee's right to privacy concerning their health information.
A) COBRA   B) SUTA   C) HIPAA   D) FICA

Employee Performance

40. The policy of _____ is a method of corrective action where consequences for improper behavior becomes more significant but gives an employee a chance to correct.
A) termination   B) HIPAA   C) progressive discipline   D) disciplinary administration

41. _____ means that either the employer or the employee may terminate employment at any time without notice or cause.
A) "At-will Employment"
B) "Right-to-Work"
C) "Fair Labor Act"
D) "ADA Act"

Chapter 9 Safety Laws and Standards

Main Topics for This Section
    A. Basic requirements of OSHA for electrical contractors
- Reporting requirements
- MSDS & Hazardous Material fact sheets

    B. Occupational Safety Laws

The chapter reviews chapter 12: Jobsite Safety and Environmental Factors of the NASCLA Contractors Guide to Business, Law, and Project Management New Jersey Edition.

Supplementary Notes

    As an electrical contractor, with few or many employees, you should be aware of the OSHA regulations for the construction industry. These regulations can be found on-line at www.osha.gov . The reference manual no longer reviews selected OSHA regulations that an electrical contractor may encounter during normal electrical projects. However, you as an employer are responsible to meet these regulations. The following are some excerpts, from the OSHA Construction Resource Manual, depicting some of the regulations you may want to become more familiar with.

- DUTY TO HAVE FALL PROTECTION –1926.501
  - Working surfaces which are 6ft or more above a lower level shall require fall protection in the form of guardrail systems, safety net systems, or personal fall arrest systems.

- SAFETY REQUIREMENTS FOR SCAFFOLDING –1910.28
  - Scaffolding shall be provide with a screen between the toe-board and the guardrail, extending along the entire opening, where persons are required to work or pass under the scaffolds.
  - The span between supports for wood planks, used on ladder / ladder jack scaffolds, shall not exceed 8 ft with a platform width of not less than 18 inches

- GENERAL REQUIREMENTS FOR STORAGE –1926.250
  - Material stored inside a building under construction shall not be placed within 6 ft of any hoistway or inside floor opening.

- THE CONTROL OF HAZARDOUS ENERGY –1910.147
  - This section requires employers to establish a program and utilize procedures for affixing appropriate lockout devices or tagout devices to energy isolating devices, to prevent unexpected energization.

OSHA Standards

1. How many parts of the OSHA standards contain the minimum requirements for the construction industry?
   A) 4   B) 3   C) 6   D) 7

2. OSHA requirements do not allow material to be stored in a building under construction within ___ from an inside floor openings.
   A) 10 ft   B) 8 ft   C) 6ft   D) 3ft

3. What is the maximum OSHA support spacing that can be used with ladder / ladder jack scaffolds utilizing wood planks?
   A) 10 ft   B) 8 ft   C) 12 ft   D) 6 ft

4. Employers must provide fall protection for employees working _____ ft or more above a working surface.
   A) 8 ft   B) 6 ft   C) 10 ft   D) 12 ft

5. Which of the following safety laws sets safety standards for construction contracts on federal projects?
   A) OSHA
   B) PEOSH
   C) Construction Safety Act
   D) OSHA Federal Act

6. OSHA regulations require employers with more than ___ employees are required to have a written emergency action plan.
   A) 12   B) 10   C) 25   D) 36

Reporting & Recording Requirements

7. Employers with more than ___ employees must also maintain records of occupational injuries.
   A) 6   B) 10   C) 11   D) 12

8. All recordable injury or illness events are recorded on which OSHA form?
   A) Form 300
   B) Form 301
   C) Form 300A
   D) All of the above

9. Which of the following accidents resulting in the stated medical treatment is **NOT** required to be recorded as a recordable occupational injury.
   A) Injury requiring application of non-prescription antiseptic
   B) Injury requiring first aid treatment only and temporary re-assignment to light duty assignment
   C) Injury requiring use of only an elastic bandage limiting the employees motion
   D) All the above

10. Which OSHA form must be completed yearly and posted summarizing the previous years' work related injuries and illnesses?
    A) 300   B) 300A   C) 301   D) 2203

11. The Form 300A must be posted by _____ and kept in place until _____.
    A) Jan. 1; June 1
    B) Feb. 1; July 1
    C) Mar. 1; Sept. 1
    D) Feb. 1; April 1

12. What is the OSHA Form required to log specific details of each recordable incident?
    A) 300   B) 300A   C) 301   D) 2203

13. OSHA records must be retained by the employer for ____ following the year to which they pertain.
    A) 1 yr   B) 2 yr   C) 5 yr   D) 7 yr

14. Records of recordable exposures must be maintained by the employer for ____.
    A) 5 years   B) 10 years   C) 15 years   D) 30 years

15. Within ____ of an occurrence of a work-related injury resulting in the death of an employee, an employer must orally notify the nearest OSHA office.
    A) 48 Hr.   B) 24 hrs   C) 4 hrs   D) 8 hrs

16. Employers do not need to report the death occurring more than ____ days after a work related incident.
    A) 7   B) 30   C) 45   D) 60

Safety Processes

17. For potentially hazardous substances an employer must have a _____ and make it available to all employees who may come in contact with this substance.
    A) First Aid Kit   B) MSDS   C) storage plan   D) Hazardous Marker Label

18. Which of the following is **NOT** required for review as part of the OSHA published inspection guidelines for enforcement of the Hazard Communications Standard.
    A) Written Hazard Communication Plan
    B) Identifying the person responsible for obtaining and maintaining MSDS
    C) Process of maintaining the required MSDS
    D) None of the Above

19. Electrical Contractors are required to contact the _____ prior to any excavation to request marking of underground utilities.
    A) Common Ground Alliance
    B) 1-800-Mark
    C) Electric Utility Co.
    D) NJ One Mark

20. NJ Law requires contractors to request utility markings at least ____days prior to commencing excavation activities.
    A) 1 full business   B) 3 full business   C) 5 full business   D) 10 full business

21. Employees have up to ____ days to notify OSHA from the time they learn of potential discriminatory action resulting from filing an OSHA complaint.
    A) 5 full business   B) 30 full business   C) 45 full business   D) 10 full business

OSHA Penalties

22. _____ is the OSHA penalty type for a violation of standards which have no direct or immediate relationship to safety and health.
    A) Other than Serious   B) Failure to Abate   C) De Minimis   D) Minor

23. _____ is the OSHA penalty type for a violation of standards that has a direct relationship to workplace safety and health but probably not result in death or serious physical harm.
    A) Other than Serious   B) Failure to Abate   C) De Minimis   D) Minor

24. Which of the following is **NOT** a general OSHA inspection type.
    A) Programmed Inspection
    B) Un-programmed Inspection
    C) Re-inspection
    D) None of the above

Environmental Organizations and Laws

25. Which federal agency is responsible for enforcement of regulations of environmental laws enacted by Congress?
    A) DEP   B) ESA   C) NPDES   D) EPA

26. Which New Jersey agency is responsible for enforcement of regulations of environmental laws?
    A) DEP   B) ESA   C) NPDES   D) EPA

27. Which department of the New Jersey DEP is responsible for encouragement, environmental stewardship, conducts site inspections and takes administration actions?
    A) Enforcement and Compliance Division
    B) Land Use Management Division
    C) Division of Administration Management
    D) Division of Water Supply

28. Which environmental law gives the EPA authority to implement water pollution control programs?
    A) The National Environmental Policy Act
    B) The National Pollution Control Act
    C) The Clean Water Act
    D) The EPA Environmental Control Act

29. Which environmental law applies to federally funded construction projects ensuring environmental impacts are reviewed during all phases of the construction project?
    A) The National Environmental Policy Act
    B) The National Pollution Control Act
    C) The Clean Water Act
    D) The EPA Environmental Control Act

30. Outdoor air quality regulations for construction activities are designed to limit the _____.
   A) vehicle emissions
   B) generation of particulate and ozone depleting substances
   C) release of CFC'c
   D) All the above

31. Contractors must submit a written Notice of Intent _____ days prior to commencement of any asbestos abatement.
   A) 5 business   B) 7 business   C) 10 business   D) 28 business

32. A renovation project involving removal of a least _____ square feet of RACM requires a written Notice of Intent to be submitted.
   A) 10   B) 60   C) 45   D) 85

33. NJ Construction sites that disturb at least ___ are required to file and obtain a NJ NPDES permit.
   A) 1   B) 0.5   C) 5   D) 3

34. What is the OSHA regulation limit of exposure (micrograms of lead per cubic meter) to lead can a worker be subjected to over an 8 hour day?
   A) 10   B) 20   C) 35   D) 50

Chapter 10 Financial Management

Main Topics for this Section

A. Accounting Methods
B. Financial Statements & Ratios
C. Accounting Methods
D. Contact Accounting
E. Earnings Recognition
F. Payment Methods
G. Payroll Accounting

The chapter reviews chapter 14: Financial Management of the NASCLA Contractors Guide to Business, Law, and Project Management New Jersey Edition.

Supplementary Notes

In this chapter the student needs to understand depreciation and payment terms computations. These topics are not well depicted but have shown up as exam questions previously. To supplement the reference manual in these areas the following examples are given of each computation.

Example 1: Straight-line Depreciation

What is the asset value of a piece of equipment, purchased for $10,000, after the 4th year of a seven year depreciation schedule?

End Of	1st yr	2nd yr	3rd yr	4th yr	5th yr	6th yr	7th yr
Asset Value	$8571.43	$7142.86	$5714.29	$4285.72	$2857.14	$1428.57	$0

Deprecation per year = Purchase Value / # of years depreciated over (depreciation schedule)
Depreciation per year = $10,000 / 7 = $1428.57

Example 2: Payment Terms

What is the payment due on a bill for a materials of $2000.00 with a 4/7 prox net 30 discount purchased and delivered on 06/10/13 if paid on 07/05/13?

- The "prox" expression specifies payment due on the date of the following month from delivery. If the bill did not have the "prox" expression the payment would have been due by June 17 in order to receive the discount and payment due in full by July 10.
- 4% discount if paid by the 7th day of the month following delivery (07/07/13)

Since payment is made on July 5th the customer receives a discount of 4% or $80 ($2000*.04) and therefore owes $1920.00.

This section also addresses payroll employee withholding and employer tax calculations such as social security and taxes. The following is an example containing several of the required calculations for various withholdings and employer tax. The IRS Circular E can be taken into the test and used for this section of question. You should down load the current copy and become familiar with it.

Example 1: Payroll Withholdings Calculation
    An employee, who is paid $1200 weekly, has filed a Form W-4 indicating Married with 3 dependants and has received only two pay checks this year, will have the following withheld from their salary;

   A. Social Security: 6.2% on the first $118,000 for calendar year 2015 wages, from Circular E 2015 edition

      Since this employee has only received two pay checks this year totaling $2400 then social security is withheld on the full portion of this pay. $1200 * .062 (6.2%) = $74.40

   B. Medicare Tax: 1.45% on all wages
      $1200 * .0145 (1.45%) = $17.40

   C. State Income Tax: Calculation not required for Business Law exam. The student only needs to know when State Tax withholdings are to be remitted to the state.

   D. Federal Tax using Circular E
      First find the Circular E tables starting in appendix A. Then based on the pay period of weekly find the WEEKLY tables . Then based on the W-4 filing of Married find the MARRIED-WEEKLY table . Then based on the W-4 again stating 3 dependants (withholding allowances) look up on the first column to find the row covering the weekly salary of $1200 and move over in the row to the column under 3 withholding allowances to obtain $106. That is the required Federal tax to withhold.

In addition to the withholdings an employer is liable for the following additional taxes on and employee's salary;

   A. Social Security: matching the employees withholding amount: $74.40
   B. Medicare Tax: matching the employee's withholding amount: $17.40
   C. Federal Unemployment Tax (FUTA): 6.2% on the first $7,000 of wages
      For this example the employee's total salary received to date is less then $7,000 therefore for this week's pay the employer will be required to pay $1200 * .062 (6.2%) = $74.40

Financial Accounting

1. A daily _____ is an event that records changes in account balances of the journals.
   A) disbursement   B) transaction   C) trial balance   D) entry

2. Which bookkeeping journal is used to record non-cash transactions?
   A) Cash Receipts and Sales journal
   B) Payroll journal
   C) General journal
   D) Cash Disbursement journal

3. Which bookkeeping journal is used to record cash payments to vendors?
   A) Cash Receipts and Sales journal
   B) Payroll journal
   C) General journal
   D) Cash Disbursement journal

4. _____ is the process of transferring the transactions recorded in the journals to the appropriate accounts.
   A) Disbursement   B) Transaction   C) Trial balance   D) Posting

5. The _____ is a numbering system that organizes the account types.
   A) chart of accounts
   B) accounting journal numbering
   C) general ledger
   D) posting

6. The financial accounts are organized in the _____.
   A) journals   B) general ledger   C) balance sheet   D) general journal

7. Under the standard chart of accounts the Revenue account is listed under which of the following numbering.
   A) 4000-4999   B) 2000-2999   C) 5000-5999   D) 7000-7999

8. The _____ is the totaling of all accounts in the general ledger.
   A) posting   B) daily transaction   C) trail balance   D) ledger balance

9. A debit to the Assets account in the general ledger will _____ the account balance?
   A) decrease   B) increase   C) balance   D) equal

10. Which of the following is not a general types of adjustment made in the Adjusted Trail Balance?
    A) prepaid expense   B) inventory adjustment   C) accrued revenue   D) earned revenue

11. Which of the following does not contribute to the balance sheet equation?
    A) Assets   B) Owners' Equity   C) Gross Profit   D) Liabilities

12. Company owned patents, franchises and goodwill are known as _____.
    A) intangible assets   B) owners' equity   C) working capital   D) long tern liability

13. Based on the balance sheet for Quality Construction (pg. 14-4 from text) what is the company's working capital?

A) $64,700   B) $18,200   C) $26,400   D) $22,300

14. The initial investment plus accumulated net profits not paid out to the owners defines the _____.

A) Current Assets   B) Owners' Equity   C) Gross Profit   D) Working Capital

15. The excess of current assets over current liabilities is a business's _____.

A) working capital   B) net worth   C) current worth   D) current income

16. Debt obligations that extend beyond one year are defined as _____.

A) Long Term Liabilities   B) Current Liabilities   C) Accrued Expense   D) Extended Expenses

17. Which financial statement summarizes the company's revenues and expenses over a given period of time?
A) Income Statement
B) Profit and Loss Statement
C) A&B
D) None of the above

18. What financial statement shows the Gross Profit of a company?
A) Statement of Income & Expense B) Balance Sheets C) Cash Flow D) General Balance

19. Expenses that are not directly linked to a particular project are accounted for under which expense account?

A) Direct Cost   B) Notes Payable   C) Accrued Expense   D) Indirect Expenses

20. The cost incurred related to marketing the business is associated to _____.

A) Long Term Liabilities   B) Current Liabilities   C) Selling Expenses   D) Extended Expenses

21. _____ is the difference between revenues and expenses.

A) Income   B) Net Profit   C) Gross Liability   D) Gross Profit

22. _____ is the difference between Income and Cost of Goods Sold.

A) Income   B) Net Profit   C) Gross Liability   D) Gross Profit

23. The Statement of Cash Flows is a financial statement that lists changes in cash based on which of the following?
A) Operating activities
B) Investing activities
C) Financing activities
D) All the above

24. The portion of the Statement of Cash Flows that measures the flow of cash between the owners and creditors is called ____.

A) Financing activities   B) Operating activities   C) Investing activities   D) Profit activities

25. Specific information related to assets and cost of pension plans or other retirement programs are documented in which of the following?
A) Income Statement
B) Profit and Loss
C) Notes to financial statements
D) Balance statement

Financial Ratios

26. Which financial ratio can be used to determine if the company is financial capable in paying it's debts?
A) Liquidity ratio   B) Quick ratio   C) Debt ratio   D) Activity ratio

27. If the liquidity ratio for a company is greater than one the company is _____.
A) in a poor position to pay its current debts
B) in a positive position to pay its current debts
C) will be unable to pay its debts
D) facing a negative cash flow

28. Which financial ratio can be used to determine how well a company effectivity manages its credit?
A) Liquidity ratio   B) Quick ratio   C) Debt ratio   D) Activity ratio

29. Which financial ratio can be used to measure the percent of total funds provided by creditors to a company?
A) Liquidity ratio   B) Quick ratio   C) Debt ratio   D) Activity ratio

30. A company's profit margin is computed from which of the following ratios?
A) Current Assets divided by Current Liabilities
B) Total Debt divided by Total Assets
C) Net Income divided by Revenues
D) Net Profit divided by Total Assets

31. A company's Return on Total Assets is computed from which of the following ratios?
A) Current Assets divided by Current Liabilities
B) Total Debt divided by Total Assets
C) Net Income divided by Revenues
D) Net Profit divided by Total Assets

Accounting Contracts and Methods

32. When a business records income and expenses as they are received and paid respectively is known as _____ method of record keeping.
A) the accrual   B) the cash   C) the simple   D) the operating cycle

33. When a business records income at the time it is earned, even though payment may not have been received is known as _____ method of record keeping.
A) the accrual   B) the cash   C) the Completed Contract   D) Percent Complete

34. What type of accounting method for a contract recognizes project revenue, costs and profits based on when the contract is finished?
A) Completed Contract Method
B) Percentage Of Completion Method
C) Percentage of Completion-Capitalization Method
D) Cash Completion Method

35. What type of accounting method for a contract recognizes project progress complete and project revenue, based on costs to date?
A) Completed Contract Method
B) Percentage Of Completion Method
C) Percentage of Completion-Capitalization Method
D) Cash Completion Method

36. What type of accounting method for a contract recognizes project progress and revenue based on cost to date after an initial minimum cost has been incurred?
A) Cost Comparison Method
B) Percentage Of Completion Method
C) Percentage of Completion-Capitalization Method
D) Cash Completion Method

37. The cumulative earnings under the Percentage of Completion method is calculated by multiplying the project percent complete by the _____.
A) Current Cost
B) Estimate Cost
C) Contract Amount
D) Estimated Profit

38. Billings in excess of cumulative earnings are considered a _____.
A) short term profit   B) current asset   C) net profit   D) current liability

39. Cumulative earnings in excess of billings are considered a_____.
A) short term profit   B) current asset   C) net profit   D) current liability

40. An invoice for $1000.00 dated July 15, 2015 NET 30 would require the bill to be paid _____.
A) within 30 business days from invoice date
B) within 30 days from invoice date
C) within 30 days from the end of the invoice month date
D) within 30 days from the 1st of the next month based on invoice date

41. When a portion of a payment is withheld until the project is completed or an approval of the work obtained is known as _____.
A) retainage   B) withholding   C) progress payment   D) depreciation

42. The Federal Prompt Payment Act requires prime contractors to receive payments within _____ after submitting progress payment invoice.
A) 5 business days   B) 10 days   C) 7 days   D) 14 days

43. The Federal Prompt Payment Act requires prime contractors to pay subcontractors within ____ after receiving payment.
    A) 5 business days   B) 10 days   C) 7 days   D) 14 days

44. The New Jersey Prompt Pay Law requires an owner to pay prime contractors not more than ____ after the billing date for approved work.
    A) 5 business days   B) 10 calendar days C) 7 business days   D) 30 calendar days

45. The New Jersey Prompt Pay Law requires an owner to approve and certify a billing within ____ after receiving it.
    A) 5 business days   B) 20 calendar days C) 7 business days   D) 14 calendar days

46. The New Jersey Prompt Pay Law contains exceptions for which of the following?
    A) Public entities
    B) Governmental entities
    C) A & B
    D) none of the above, no exceptions exist

47. Unless agreed to in writing, the New Jersey Prompt Pay Law requires a prime contractor to pay the subcontractor due payments within _____.
    A) 5 business days   B) 10 calendar days C) 7 business days   D) 30 calendar days

48. Unless agreed to in writing, the New Jersey Prompt Pay Law requires a subcontractor to pay their subcontractor within _____.
    A) 5 business days   B) 10 calendar days C) 7 business days   D) 30 calendar days

49. The New Jersey Prompt Pay Law allows for the owed party to charge interest on late payments at prime rate plus _____ percent.
    A) 1   B) 2   C) 3   D) 5

50. The New Jersey Prompt Pay Law allows the owed party to charge interest on late payments only after providing _____ written notice to the party required to make payment.
    A) 5 business days   B) 10 business days C) 7 calendar days   D) 30 calendar days

51. What is the asset value of a tool purchased for $2000.00 at the end of the 3rd year of a 5 yr schedule using straight line depreciation?
    A) $800   B) $400   C) $666.66   D) $1200

52. Which of the following is **NOT** a method for depreciation?
    A) MACRS   B) Straight-Line   C) Accelerated   D) Half Year

53. Which of the following, when listed on a shipping invoice, indicates shipping has been paid by seller?
    A) FOB Freight Prepaid
    B) FAS Freight Prepaid
    C) FAS port storage
    D) FOB Freight Allowed

54. How much is owed on a bill for a materials of $1000.00 with a 3/10 net 30 on invoice, purchased and shipped on 01/02/15, if paid on 01/11/15?
   A) $1000.00   B) $900.00   C) $970.00   D) $980.00

55. How much is owed on a bill for a materials of $1000.00 with a 3/10 EOM on invoice, purchased and shipped on 01/16/15, if paid on 02/05/14?
   A) $1000.00   B) $900.00   C) $1030.00   D) $970.00

56. How much is owed on a bill for a materials of $1000.00 with a COD on invoice, purchased and shipped on 01/16/15, if delivered on 02/05/14?
   A) $1000.00   B) $900.00   C) $1030.00   D) $970.00

57. As an employer, how much tax must be withheld for Medicare from an employee's salary?
   A) 6.2%   B) 2.9%   C) 1.45%   D) none

58. As an employer, you pay matching _____ tax for each employee.
   A) Federal & Social Security
   B) Medicare & Social Security
   C) Medicare & Federal
   D) FUTA & Social Security

59. Social Security must be withheld on _____ the employee's salary for 2015 wage base.
   A) the first $90,000 of
   B) the first $133,700 of
   C) all of
   D) the first $118,000 of

60. As an employer, what is the total Social Security tax withheld on an employee with a gross annual (2015) salary of $120,000.
   A) $7316   B) $7440.00   C) $14632.00   D) $10899.60

61. As an employer, what is the total Social Security tax due on an employee with a gross (2015) annual salary of $120,000.
   A) $7316   B) $7440.00   C) $14632.00   D) $14880.00

62. How much federal tax table on page 14-14 text must be withheld from a weekly salary of $1,100, for an employee who has filed a W-4 with married and for 3 withholding allowances?
   A) $122.00   B) none   C) $87.00   D) $89.00

63. On an employee's second weekly salary of $1000, what is the employers total Social Security, Medicare and FUTA tax liability on this salary?
   A) $138.50   B) $76.50   C) $215.00   D) $277.00

64. As an employer, when determining total wages to compute your Social Security, FUTA and Medicare tax, which of the following is NOT part of the total wages paid?
   A) Personal use of the company truck
   B) Sick pay
   C) Vacation pay
   D) Your cost of an employee's health insurance

Chapter 11 Tax Basics

Main Topics for this Section
   A. Payroll tax
   B. Use of Circular E, Employer's Tax Guide
   C. New Jersey Sales Tax

The chapter reviews chapter 15: Tax Basics of the NASCLA Contractors Guide to Business, Law, and Project Management New Jersey Edition.

Supplementary Notes

This section addresses payroll employee withholding and employer tax calculations such as social security and taxes. The following is an example containing several of the required calculations for various withholdings and employer tax. The IRS Circular E can be taken into the test and used for this section of question. You should down load the current copy and become familiar with it.

Example 1: Payroll Withholdings Calculation
   An employee, who is paid $1200 weekly, has filed a Form W-4 indicating Married with 3 dependants and has received only two pay checks this year, will have the following withheld from their salary;

   A. Social Security: 6.2% on the $118,000 for calendar year 2015 wages, from Circular E 2013 edition

   Since this employee has only received two pay checks this year totaling $2400 then social security is withheld on the full portion of this pay. $1200 * .062 (6.2%) = $74.40

   B. Medicare Tax: 1.45% on all wages
   $1200 * .0145 (1.45%) = $17.40

   C. State Income Tax: Calculation not required for Business Law exam. The student only needs to know when State Tax withholdings are to be remitted to the state.

   D. Federal Tax using Circular E
   First find the Circular E tables starting in appendix A. Then based on the pay period of weekly find the WEEKLY tables . Then based on the W-4 filing of Married find the MARRIED-WEEKLY table . Then based on the W-4 again stating 3 dependants (withholding allowances) look up on the first column to find the row covering the weekly salary of $1200 and move over in the row to the column under 3 withholding allowances to obtain $106. That is the required Federal tax to withhold.

In addition to the withholdings an employer is liable for the following additional taxes on and employee's salary;

A. Social Security: matching the employees withholding amount: $74.40
B. Medicare Tax: matching the employee's withholding amount: $17.40
C. Federal Unemployment Tax (FUTA): 6.2% on the first $7,000 of wages
   For this example the employee's total salary received to date is less then $7,000 therefore for this week's pay the employer will be required to pay $1200 * .062 (6.2%) = $74.40

The following section covers New Jersey State sales tax. This is not covered in detail within the N.J. Electrician's Reference manual.

Electrical Contractor's providing services with in the state of New Jersey are required to collect 6% sales tax and remit it to the state. The following is a re-print from the New Jersey Division of Taxation summarizing when a contractor is required to collect and remit sale tax.

"   A contractor is anyone in the business of working on the real property of others. The term "contractor" also includes those who manufacture, sell, and install items, which become part of real property. These contractors are known as fabricator/contractors. The work performed by a contractor can be a capital improvement, a repair, or maintenance.

Contractors working in New Jersey are required to register with the State of New Jersey and collect Jersey sales tax on charges for labor. Contractors are required to pay sales tax on the materials supplies, equipment and services they purchase, rent or use, when performing work on the real property of others. When a contractor performs work that results in a capital improvement, no sales tax is charged on the labor portion of the bill if the property owner supplies a properly completed Certificate of Capital Improvement, Form ST-8. The contractor must keep a copy of the form on file to document why no sales tax was collected on the labor portion of the bill. When a contractor performs repair or maintenance service, sales tax must be charged on the labor portion of the bill. If the labor and materia charges are not separately stated on a bill, the contractor must collect sales tax on the total amount of the bill.

A subcontractor is a contractor who performs specific tasks for a prime contractor. Purchases o materials by a subcontractor are subject to sales tax in the same manner as purchases made by a prime contractor. A subcontractor who performs taxable services for a prime contractor does not charg prime contractor sales tax on the labor portion of the bill. The prime contractor is responsible for collecting any sales tax due from the property owner for work performed by the subcontractor.

For more information about contractors request a copy of the Tax Topic Bulletin, S&U-3, Contractors and New Jersey Taxes. "

Federal Taxes

1. Prior to hiring employees and withholding payroll taxes you are required to obtain _____.
   A) FEIN   B) Social Security Number   C) I-9   D) W4

2. Which of the following entities does **NOT** require a Federal Employer Identification Number?
   A) LLC with one employee
   B) Partnership
   C) Sole Proprietorship with no employees
   D) All the above

3. Which of the following is a type of business tax?
   A) Income Tax
   B) Self-employment tax
   C) Employment taxes
   D) All the above

4. All business entities except _____ must file an annual income tax return.
   A) Sole Proprietor   B) Partnership   C) S-Corporation   D) Corporation

5. Which of the following has to make estimated tax payments if expected to owed tax is $800?
   A) Sole Proprietor   B) Partnership   C) S-Corporation   D) Corporation

6. Which of the tax is required only by an individual who works for themselves?
   A) Income Tax
   B) Self-employment tax
   C) Employment taxes
   D) All the above

7. Employment tax records must be retained for at least ____ years.
   A) 2   B) 3   C) 4   D) 10

8. As an employer, you pay matching _____ tax for each employee.
   A) Federal & Social Security
   B) Medicare & Social Security
   C) Medicare & Federal
   D) FUTA & Social Security

9. Social Security must be withheld on _____ the employee's salary for 2015 wage base.
   A) the first $90,000 of
   B) the first $118,000 of
   C) all of
   D) the first $113,700 of

10. What form must be filed by the employee to determine the Employee's Withholding Allowance?
    A) 1090   B) W-2   C) W-4   D) W-1040.

11. As an employer, how much tax must be withheld for Medicare from an employee's salary?
    A) 6.2%   B) 2.9%   C) 1.45%   D) none

12. As an employer, what is the total Social Security tax withheld on an employee with a gross annual (2015) salary of $120,000.
    A) $7316.00   B) $7440.00   C) $10788.00   D) $14880.00

13. As an employer, what is the total Social Security tax due on an employee with a gross (2015) annual salary of $120,000?
    A) $7316.00   B) $7440.00   C) $14098.80   D) $14880.00

14. What form must be sent to each employee summarizing the employee's previous year's wages?
    A) W-2   B) 1096   C) 1099   D) 10W40

15. If your tax withholding liability for the previous four quarters was less than $50,000 than your required depositor schedule is _____.
    A) monthly   B) semiweekly   C) weekly   D) daily

16. If your tax withholding liability for the previous four quarters was greater than $50,000 than your required depositor schedule is _____.
    A) monthly   B) semiweekly   C) weekly   D) daily

17. Your tax withholding depositor schedule for the first year of operation is _____ .
    A) monthly   B) semiweekly   C) weekly   D) daily

18. If you report less than _____ in tax withholding for a quarter you can use the IRS Form 941 Quarterly Employer's Tax Return to make payments.
    A) $500   B) $1000   C) $1500   D) $2500

19. As an employer, how much FUTA tax will be withheld from an employee with a salary of $75,000?
    A) $434.00   B) $1052.70   C) $1087.50   D) none

20. As an employer, what is the FUTA tax liability due from an employee with a salary of $85,000?
    A) $434.00   B) $4724.40   C) $1087.50   D) none

21. If your FUTA tax liability is less than _____ you are not required to make a deposit but may elect to carry forward to the next quarter.
    A) $100   B) $250   C) $500   D) $750

22. As an employer, if your FUTA is deposited quarterly and you have a total of $990 for deposit on March 31, 20XX, by what date are you required to filed / deposited this tax to the government?
    A) March 31, 20XX   B) April 30, 20XX   C) June 30, 20XX   D) July 31, 20XX

23. As an employer what form must you file to remit / deposit FUTA tax?
    A) Form 940   B) Form 1040   C) Form 1080   D) Form 1120

24. A _____ of the tax liability penalty is imposed on an employer for failure to pay "trust fund" taxes, defined as withheld taxes.
    A) 2%   B) 100%   C) 10%   D) 5%

25. A _____ of the tax liability penalty is imposed on an employer for "trust fund" tax payments, defined as withheld taxes, paid 10 days late.
   A) 2%   B) 100%   C) 10%   D) 5%

26. What form must be sent to each employee / consultant not subject to withholding?
   A) W-2   B) 1096   C) 1099   D) 10W40

27. What form must be sent to the IRS summarizing the wages / compensation paid to each employee / consultant not subject to withholding?
   A) 1099   B) 1096   C) W-2   D) 1090

28. A new W-4 form is required to be completed before _____ of each year for all employees claiming a tax withholding exemption.
   A) Jan 31   B) Feb 15   C) Dec 1   D) Mar 31

29. An employer's Form 1096 must be filed with the IRS by _____.
   A) Jan 31   B) Feb 15   C) Feb 28   D) Mar 31

New Jersey State Taxes

30. Corporations doing business in New Jersey are required to complete a _____ tax return.
   A) CBT-100 Corporate Business
   B) C-1040 Corporate Business
   C) Form 1058 NJ Business
   D) Form 20S Business

31. S-Corporations doing business in New Jersey are required to complete a _____ tax return.
   A) CBT-100 Corporate Business
   B) C-1040 Corporate Business
   C) Form 1058 NJ S-Corporation
   D) Form 20S S-Corporate

32. A corporation doing business in NJ that is required to pay estimated taxes will pay ___ of the current estimated tax liability in the 6th month of the accounting period.
   A) 25%   B) 30%   C) 50%   D) 60%

33. LLC's doing business in NJ are required to file a _____ tax return.
   A) Form-1065   B) Form 1099   C) Form 1040   D) Form CBT-100

34. Withheld employee state income tax must be reported and submitted by the _____ day of the month following the end of the quarter when filing on a quarterly schedule.
   A) 1st   B) 15th   C) 20th   D) 30th

35. Withheld employee state income tax are required to file Form _____ Employer Report of Wages Paid by the 30th day of the month following the end of the quarter.
   A) WR-1040   B) WR-30   C) WR-4   D) WR-1058

New Jersey Sales and Use Tax

36. How much sales tax must be collected on a taxable sale of $250.00?
    A) $6   B) $12   C) $15   D) $16.25

37. How much sales tax must be collected on a bill totaling $650.00 for a service panel replacement where the bill states in writing charges billed 1. Material : $200 ; 2. Labor: $450 ?
    A) $12   B) $27   C) $39   D) none required

38. How much NJ sales tax must be charge for $100 of materials purchased from a supplier outside NJ were no sales tax is charged?
    B) $6   B) $7   C) $13   D) none required

Chapter 12 NJ Construction Lien Laws

Main Topics For This Section
  A. What is a Construction Lien
  B. Filing a lien
  C. Priority of a Lien

The chapter reviews chapter 16: New Jersey Construction Lien Law and Appendix J: NJ Statue 2A:44 of the NASCLA Contractors Guide to Business, Law, and Project Management New Jersey Edition.

Supplementary Notes

This section describes a significant method to secure payment in the event of non-payment from the other party involved in the contract. However as advised in the reference manual an attorney should be contacted for assistance in the filing. Use this section as a reference for determining when a lien needs or must be filed.

New Jersey Construction Lien Law

1. What property is **NOT** subject to the New Jersey Construction Lien Law?
   A) Residential property for which the owner contracted the work
   B) Commercial property for which the leasing agent contracted the work
   C) Public property
   D) Commercial property for which the General Contractor of the project contracted the work

2. A non-residential lien must be filed no later than _____ from the date following the last work, services or material was provided.
   A) 90 days   B) 30 days   C) 1 year   D) 60 days

3. To file a lien the contractor must file it with _____.
   A) County Clerk of the county the contractor's business is registered in.
   B) County Clerk of the county containing the property for which the lien is for.
   C) Superior Court of the county the contractor's business is registered in.
   D) Superior Court of the county containing the property for which the lien is for.

4. Within _____ days following the filing of the lien, the claimant shall notify the owner of the property.
   A) 10 calendar   B) 30 calendar   C) 10 business   D) 30 business

5. A lien filed against an owner by an electrical contractor shall be paid out of the lien fund in what order?
   A) Mortgage holder first then split equally among all other lien holders.
   B) Based on when the lien was filed. (1st filed first paid)
   C) Labor wages first, sub-contractors then the general contractor.
   D) Equal with all other liens.

6. A claimant has _____ to file a certification to discharge lien claim record or maybe liable for cost incurred by the owner to discharge the lien.
   A) 90 days
   B) within 30 days
   C) within 10 business days
   D) 45 days

7. A claimant filing a lien claim shall forfeit all rights to enforce the lien if the claimant fails to commenc an action in the Superior Court, in the county in which the real property is situated, within ____ of th date of last provision of work or services.
   A) 90 days
   B) 30 days
   C) 1 year
   D) 6 months

Chapter 13 Business Fundamentals

Main Topics For This Section
  A. What is a Construction Lien
  B. Filing a lien
  C. Priority of a Lien

The chapter reviews several miscellaneous chapters of the NASCLA Contractors Guide to Business, Law, and Project Management New Jersey Edition. These sections contain primarily general information on operating a business or direction to other areas and websites to obtain additional information and contain little specific information generally found on a standardized test. Sections of the NASCLA Contractors Guide to Business, Law, and Project Management New Jersey Edition covered in this review chapter include Chapter 5: Your Business Toolbox; Chapter 6 Marketing and Sales; Chapter 10 Customer Relations and Chapter 13: Working with Subcontractors.

Supplementary Notes

   None

Your Business Toolbox

1. Establishing a _____ is the first step to ensuring that good ethics are practiced throughout your company.
   A) Business Plan
   B) OSHA Plan
   C) Code of Conduct
   D) Diversity Plan

2. Which organization can provide counseling, assistance to small business owners?
   A) SBA   B) SAB   C) SCOR   D) SBC

3. Which organization can provide counseling, assistance, and training from retired executives to small business owners?
   A) SBA   B) SAB   C) SCORE   D) SBC

4. Information on NJ state level Small Business Certifications can be obtain from _____.
   A) The U.S Small Business Administration
   B) County Clerk Office
   C) 8(a) Business Development Office
   D) NJ Dept. of the Treasury

Marketing and Sales

5. A _____ is a formal document outlining the company's vision, customer base and methods for promotion and industry opportunities and challenges.
   A) Business Plan   B) Marketing Plan   C) Risk Mitigation Plan   D) MSDS

6. The process of advertising, promotions, pricing strategies and timely distribution are known as _____.
   A) Sales   B) Marketing   C) Risk Mitigation   D) Business Planning

Customer Relations

7. Which of the following would **NOT** be part of a negotiation of final contract agreement for a project?
   A) Price of work
   B) Completion date
   C) Warranties
   D) None of the above

Working with Subcontractors

8. What are the three categories that must be examined to determine the IRS status between an employee or independent contractor?
   A) Financial control, Ownership, Risk
   B) Behavioral control, Financial control, Risk
   C) Type of Relationship between Parties, Risk, Financial control
   D) Behavioral control, Financial control, Type of Relationship between Parties

9. What IRS form must be completed to request a review and determination of a workers status for purposes of Federal Employment taxes?
   A) 1099   B) SS-8   C) W-4   D) S-100

Chapter 14 Practice Exams

This section contains two practice exams. Each exam should be completed in two hours. Tear out the bubble answer sheet and place all answers here just as you would on the actual exam. This wil give you some test taking practice. You should review the test taking tips in the Introduction section and pace yourself. Once you have completed each test check your answers against those in section 13. You must get 35 correct (no more than 15 wrong) to achieve a passing grade (70%).

Practice Exam #1

Score_____

1. A B C D	26. A B C D
2. A B C D	27. A B C D
3. A B C D	28. A B C D
4. A B C D	29. A B C D
5. A B C D	30. A B C D
6. A B C D	31. A B C D
7. A B C D	32. A B C D
8. A B C D	33. A B C D
9. A B C D	34. A B C D
10. A B C D	35. A B C D
11. A B C D	36. A B C D
12. A B C D	37. A B C D
13. A B C D	38. A B C D
14. A B C D	39. A B C D
15. A B C D	40. A B C D
16. A B C D	41. A B C D
17. A B C D	42. A B C D
18. A B C D	43. A B C D
19. A B C D	44. A B C D
20. A B C D	45. A B C D
21. A B C D	46. A B C D
22. A B C D	47. A B C D
23. A B C D	48. A B C D
24. A B C D	49. A B C D
25. A B C D	50. A B C D

Practice Exam #2

Score_____

1. A B C D	26. A B C D
2. A B C D	27. A B C D
3. A B C D	28. A B C D
4. A B C D	29. A B C D
5. A B C D	30. A B C D
6. A B C D	31. A B C D
7. A B C D	32. A B C D
8. A B C D	33. A B C D
9. A B C D	34. A B C D
10. A B C D	35. A B C D
11. A B C D	36. A B C D
12. A B C D	37. A B C D
13. A B C D	38. A B C D
14. A B C D	39. A B C D
15. A B C D	40. A B C D
16. A B C D	41. A B C D
17. A B C D	42. A B C D
18. A B C D	43. A B C D
19. A B C D	44. A B C D
20. A B C D	45. A B C D
21. A B C D	46. A B C D
22. A B C D	47. A B C D
23. A B C D	48. A B C D
24. A B C D	49. A B C D
25. A B C D	50. A B C D

Practice Exam #1

1. Which of the following is not a key function of a business plan?

   A) Financial Books   B) Planning Tool   C) Loan Document   D) Benchmarking Tool

2. What is the simplest and least expensive business type to setup and operate?

   A) LLC
   B) Partnership
   C) Individual Unregistered
   D) Sole Proprietorship

3. Profits earned from revenue in a sole proprietorship are taxed as_____.

   A) Personal income   B) S-Corp income   C) Corporate income   D) Business income

4. Which element of a business plan contains the uniqueness of your product or service as well as your pricing, adverting and promotional strategies?

   A) Product & Services   B) Company summary   C) Market analysis   D) Market strategy

5. Which type of business organization does **NOT** allow for avoidance of double taxation on earnings?

   A) S-Corporation   B) LLC   C) C-Corporation   D) B&C

6. Which of the following is **NOT** exempt electrical work.

   A) Replacement of a 277V lamp
   B) Replacement of a 120V fuses
   C) Installation of Transmission Lines under contract for the local public utility
   D) Employed qualified journeyman electrician for the school district installing 120V light fixtures

7. After a business's registration is filed with the State Treasurer's office a domestic business name will be reserved for _____.

   A) 120 days   B) 90 days   C) 60 days   D) 30 days

8. How many public members not associated with the electrical industry sit on the N. J. Board of Examiners?

   A) 3   B) 7   C) 2   D) 1

9. How long is an appointment on the N. J. Board of Examiners?

   A) 1 yr.   B) 3 yrs   C) 5 yrs   D) 4 yrs

10. On a commercial vehicle utilized in the practice of electrical contracting the required identification information shall visible mark on the vehicle with lettering at least_____ height.

   A) 2 in   B) 3 in   C) 4 in   D) 5 in

11. A _____ insurance policy protects the contractor against any physical loss or damage to the project or project materials, except for listed exclusions.

   A) Liability   B) All-Risk Builder's Risk   C) Named Peril Builder's Risk   D) Equipment Floater

12. If changes are issued to a bid package prior to the bid submittal date a(n) _____ is issued officially and becomes part of the bid package?

   B) Supplement   B) bid form   C) addendum   D) contract modification

13. When available space for identification lettering is limited by design of the vehicle and making strict compliance with marking letter size, the lettering size shall be reduced to _____.

   A) 1 in
   B) 2 in
   C) 2 1/2 in
   D) as close as possible to 3 in

14. Which of the following is required n order for a contract to be enforceable?.

   B) "Consideration" B) Completion Dates C) Specific Description of the work D) Owner's name

15. Of all continuing education credits required, what is the minimum number required that must be related to the applicable State statutes and rules?

   A) 1   B) 3   C) 6   D) 9

16. As an electrical contractor you are required to provide constant on-site supervision for electricians employed with less than _____ experience working under the UCC.

   A) 3 ½ yrs   B) 2 ½ yrs   C) 1 ½ yrs   D) 1 yr

17. The withholding of a percentage of payment from the contractor until final completion and acceptance of the work is known as _____?

   A) a warranty   B) retainage   C) indemnification   D) limited liability

18. Any person violating the Professional and Occupation Uniform Enforcement act shall be guilty of.....

   E) a 4th degree felony.
   F) a crime with the severity based on the Boards findings.
   G) a misdemeanor.
   H) Misconduct.

19. A _____ is a "Breach of Contract" which may result in a termination of the contract or the injured party suing for only the value of the damages incurred.

A) Immaterial Breach   B) Material Breach   C) Indirect Breach   D) Contractual Liquidation

20. If an applicant fails to obtain a passing score on their third attempt they shall_____

A) be required to retake ALL exams over.
B) be required to wait 1 yr prior to retake the failed sections.
C) be required to wait 6 months prior to retaking the failed section.
D) be required to re-apply to the Board for approval to retake exam.

21. Calculating the _____ ranges for each task provides a guide to determine how on schedule task or the entire project may be.

A) activity duration   B) labor charge   C) earliest & latest time   D) float time

22. What of the following federal law requires reporting of new hires to the State?

A) The Federal Reporting Work Act of 1996
B) The NJ New Hire Reporting Program
C) The Work and Opportunity Reporting Act of 1996
D) The Personal Responsibility and Work Opportunity Reconciliation Act

23. Unemployment Compensation insurance tax must be paid by the employer if they had paid wages in excess of _____ within any quarter of the year.

A) $1000   B) $1500   C) $2500   D) $5000

24. For a licensed well driller / installer which of the following electrical task can he **NOT** perform?

A) Installation of 240V, 1ϕ exterior underground wiring from the controller to the well pump
B) Installation of 6ft of 120V, 1ϕ interior wiring from the OCP device to the pump disconnect
C) Installation of 8ft of 12-2 Romex run inside to connect the pump to the pressure switch
D) Installation of 240V, 1ϕ exterior branch circuit wiring to an exterior controller

25. How many parts of the OSHA standards contain the minimum requirements for the construction industry?

A) 4   B) 3   C) 6   D) 7

26. _____ outlines the stages of construction with anticipated revenues and expenditures.

A) Working Capital schedule
B) Cash flow budget
C) Earn Value
D) Cash management

27. All recordable injury or illness events are recorded on which OSHA form?

   A) Form 300
   B) Form 301
   C) Form 300A
   D) All of the above

28. A claim under a _____ bond need not be specifically outlined in the bond.

   A) private
   B) common-law
   C) statutory
   D) N/A all bonds address only provisions written into the bond

29. Which bookkeeping journal is used to record non-cash transactions?

   A) Cash Receipts and Sales journal
   B) Payroll journal
   C) General journal
   D) Cash Disbursement journal

30. Cumulative earnings in excess of billings are considered a_____.

   A) short term profit   B) current asset   C) net profit   D) current liability

31. The Miller Act requires a payment bond in the sum of _____ for a $2M project.

   A) equal to Performance Bond
   B) not to be greater then $2.5 million dollars
   C) $800,000 dollars
   D) $2,000,000 dollars

32. Which of the following entities does **NOT** require a Federal Employer Identification Number?

   A) LLC with one employee
   B) Partnership
   C) Sole Proprietorship with no employees
   D) All the above

33. What type of contract relationship has the contractor being responsible for the entire design, construction process and all financial and land acquisition responsibilities?

   A) Design to Cost   B) Fast Track Design   C) Design / Build   D) Turn Key

34. What is the asset value of a tool purchased for $2000.00 at the end of the 3rd year of a 5 yr schedule using straight line depreciation?

   A) $800   B) $400   C) $666.66   D) $1200

35. _____ insurance provides coverage over and above Workers' Compensation in case of judgments stemming from civil actions due to injury or death of an employee.

    A) Contractual Liability Insurance
    B) Employee Operations Liability
    C) Contractor's Protective Public and Property Damage Liability
    D) Employer's Liability

36. Your tax withholding depositor schedule for the first year of operation is _____ .

    A) monthly   B) semiweekly   C) weekly   D) daily

37. A non-residential lien must be filed no later than _____ from the date following the last work, services or material was provided.

    A) 90 days   B) 30 days   C) 1 year   D) 60 days

38. Conduit delivery costs are included under what part of the estimate?

    A) Material Cost   B) Labor Cost   C) Equipment Cost   D) Subcontractors

39. As an employer, how much tax must be withheld for Medicare from an employee's salary?

    A) 6.2%   B) 2.9%   C) 1.45%   D) none

40. If your estimate included $3000 for direct cost and $2500 for indirect cost with a Profit margin of 10% of total project cost, what would be the total bid price submitted?

    A) $550   B) $5500   C) $6050   D) $5800

41. What type of accounting method for a contract recognizes project progress complete and project revenue, based on costs to date?

    A) Completed Contract Method
    B) Percentage Of Completion Method
    C) Percentage of Completion-Capitalization Method
    D) Cash Completion Method

42. Within _____ days following the filing of the lien, the claimant shall notify the owner of the property

    A) 10 calendar   B) 30 calendar   C) 10 business   D) 30 business

43. What type of insurance is mandatory for a contractor who employs employees and benefits are accrued through non-voluntary employee and employer contributions?

    A) Social Security
    B) Workers Compensation
    C) Employee's liability
    D) FUTA

44. _____ is the individual responsible for the management of daily onsite operations of the project.

   A) Forman   B) Project Manager   C) Superintendent   D) Expediter

45. What project document section defines changes to the released bid package issued prior to bid?

   A) Supplementary Conditions   B) Supplements   C) Allowances   D) Addenda

46. Which of the following ADR processes uses a third party to act as a judge and render a legally binding decision to resolve the dispute?

   A) Mediation   B) Negotiation   C) Collaborative Law   D) Arbitration

47. What is the production rate (# outlets installed/hour) for a job that had 3 electricians rough-in a residential unit (service not included), which had 88 electrical outlets, over 3 (8-hour) day period?

   A) 3.7   B) 29.2   C) 1.22   D) 2.33

48. A company's profit margin is computed from which of the following ratios?

   A) Current Assets divided by Current Liabilities
   B) Total Debt divided by Total Assets
   C) Net Income divided by Revenues
   D) Net Profit divided by Total Assets

49. A contractor may develop _____ to detail specific aspects of a project.

   A) Purchase Order   B) Contract Specifications   C) Shop Drawings D) Blue Prints

50. NJ employers who fail to submit new hire information as required to the NJ New Hire Reporting Program shall be subject to a fine up to ____ per incident.

   A) $25   B) $500   C) $300   D) $2500

Practice Exam #2

1. A claimant has _____ to file a certification to discharge lien claim record or maybe liable for cost incurred by the owner to discharge the lien.

   A) 90 days
   B) within 30 days
   C) within 10 business days
   D) 45 days

2. _____ is the process of transferring the transactions recorded in the journals to the appropriate accounts.

   A) Disbursement   B) Transaction   C) Trial balance   D) Posting

3. NJ employers who submit hiring reports magnetically or electronically, to the NJ New Hire Reporting Program, must submit the transmissions ____monthly and no longer than _____ days apart.

   A) Twice - 5   B) Four times - 7   C) Twice - 16   D) Twice - 30

4. To clarify an architect's plans and specifications _____ maybe required to be produced by the material supplier or contractor.

   A) Work Specs   B) Cut Sheet   C) Contracts Letter of Clarification   D) shop drawings

5. An invoice for $1000.00 dated July 15, 2015 NET 30 would require the bill to be paid _____.

   A) within 30 business days from invoice date
   B) within 30 days from invoice date
   C) within 30 days from the end of the invoice month date
   D) within 30 days from the 1st of the next month based on invoice date

6. Which of the following ADR processes is a facilitative process where parties agree to resolve the dispute without litigation thru discussions between their advocates?

   A) Mediation   B) Negotiation   C) Collaborative Law   D) Arbitration

7. Legal fees related to the bid and contracts are charged as?

   A) Company Overhead   B) Direct Labor Cost   C) Indirect Labor Cost   D) Project Overhead

8. What type of accounting method for a contract recognizes project progress and revenue based on cost to date after an initial minimum cost has been incurred?

   A) Cost Comparison Method
   B) Percentage Of Completion Method
   C) Percentage of Completion-Capitalization Method
   D) Cash Completion Method

9. The Miller Act requires a payment bond in the sum of _____ for a $2M project.

   A) equal to Performance Bond
   B) not to be greater than $2.5 million dollars
   C) $800,000 dollars
   D) $2,000,000 dollars

10. Which element of a business plan contains the uniqueness of your product or service as well as your pricing, adverting and promotional strategies?

   A) Product & Services   B) Company summary   C) Market analysis   D) Market strategy

11. What is the method of estimating called when the estimate is based on a listing all materials and labor needed for the project and assigning a specific price to each?

   A) Quantity Take-off Method B) Unit Price Method C) Approximation Method D) SWAG Method

12. _____ is the individual responsible for timely delivery of all materials work on the project.

   A) Forman   B) Project Manager   C) Architect   D) Expediter

13. LLC's are taxed based on _____.

   A) the business entity
   B) the number of business owners
   C) a sole proprietorship
   D) DBA filing

14. Business income and sales taxes are the responsibility of _____.

   A) General partners only
   B) limited for Limited partners
   C) all partners
   D) stock holders

15. Which of the following has to make estimated tax payments if expected to owed tax is $800?

   A) Sole Proprietor   B) Partnership   C) S-Corporation   D) Corporation

16. _____ is when a subcontractor approaches the general contractor post bid award with a revised lower bid in an attempt to capture the bid award.

   A) Bid Shopping   B) Bid Peddling   C) Bid Rigging   D) Low Bid Marketing

17. Which of the following scheduling methods indicate the interdependencies of the activities.

   A) Calendar Scheduling
   B) Bar Chart
   C) Critical Path Method
   D) All the above

18. Under the standard chart of accounts the Revenue account is listed under which of the following numbering.

A) 4000-4999   B) 2000-2999   C) 5000-5999   D) 7000-7999

19. What is it that must be developed to start a successful business that will define how you will meet your primary business goals?

A) Financial Plan   B) Goals   C) Business Plan   D) Business Organization

20. After a business's registration is filed with the State Treasurer's office a domestic business name will be reserved for _____.

A) 120 days   B) 90 days   C) 60 days   D) 30 days

21. As an employer, you pay matching _____ tax for each employee.

A) Federal & Social Security
B) Medicare & Social Security
C) Medicare & Federal
D) FUTA & Social Security

22. When a contract imposes monetary damages for failure to meet specific provisions of the contract, these damages are known as _____.

A) Performance Penalties
B) Liquidated Damages
C) Breach of Contract
D) Performance Forfeitures

23. How long must an applicant wait before he/she is eligible to re-take failed sections of the licensing examination?

A) 3 months   B) 6 months   C) 1 yr.   D) 6 weeks

24. Employers must complete an I-9 form with required documentation within ____ days from the date of employment.

A) 3   B) 7   C) 10   D) 22

25. Which of the following is **NOT** a method of obtain approved continuing education credits?

A) CE class pending Board approval
B) Authorship of an 8000 word textbook related to the practice of electrical contracting
C) Authorship of a published article relate to electrical contracting with only 250 words in length
D) All of the above

26. _____ is a bond that provides coverage against defects in workmanship or materials for a stated period.

A) Maintenance Bond   B) Warranty Bond   C) Performance Bond   D) Lien Bond

27. If your tax withholding liability for the previous four quarters was less than $50,000 than your required depositor schedule is _____.

A) monthly   B) semiweekly   C) weekly   D) daily

28. Any person violating the Professional and Occupation Uniform Enforcement act shall be liable to a civil penalty of not more than _____ for the first violation.
   A) $5000
   B) $7500
   C) $10000
   D) $20000

29. _____ insurance provides protection when a contractor is liable for the acts of others for whom he has responsibility.

   A) Contractual Liability
   B) Completed Operations Liability
   C) Contractor's Protective Public and Property Damage Liability
   D) Professional Liability

30. Domestic and foreign corporations are required to _____ before filing Articles of Incorporation.
   A) register with NJ Division of Revenue
   B) reserve a corporation name
   C) file a Form 2553
   D) file for distribution of dividends

31. A "Qualified Journeyman" has acquired _____ of practical experience and ___ classroom hours of related instruction.

   A) 6 months / 2 years   B) 4000 hrs / 544 hrs   C) 5 years / 2 yrs   D) 8000 hrs / 576 hrs

32. As an electrical contractor you are required to provide on-site supervision for electricians employed and can employ a qualified journeyman to perform this task provided _____.
   A) the journeyman is on-site providing constant supervision for electricians with less than 3 ½ yrs experience
   B) the journeyman provides directions and inspects work of electricians with more than 3 ½ yrs experience
   C) A & B
   D) they have a license (Journeyman cannot be delegated supervisory requirement)

33. What type of contract relationship has the contractor beginning the construction prior to the completion of contract documents?

   A) Design to Cost   B) Fast Track Design   C) Design / Build   D) Turn Key

34. The initial investment plus accumulated net profits not paid out to the owners defines the _____.

   A) Current Assets   B) Owners' Equity   C) Gross Profit   D) Working Capital

35. Which of the following safety laws sets safety standards for construction contracts on federal projects?

A) OSHA
B) PEOSH
C) Construction Safety Act
D) OSHA Federal Act

36. Unemployment Compensation insurance tax must be paid by the employer if they had paid wages in excess of _____ within any quarter of the year.

A) $1000   B) $1500   C) $2500   D) $5000

37. Which of the following is **NOT** covered by social security insurance?

A) retirement benefits   B) survivor benefits   C) long term disability benefits   D) medical benefits

38. A licensed electrical contractor must receive, at a minimum, _____ hours of continuing education related to the most recent edition of the NFPA-70 for each triennial registration period.

A) 9   B) 10   C) 24   D) 34

39. _____ is the difference between revenues and expenses.

A) Income   B) Net Profit   C) Gross Liability   D) Gross Profit

40. For a business who's qualified licensee is rendered incapable of fulfilling their professional duties the may grant a delay in returning the licensee's for a maximum of _____

A) 30 days   B) 90 days   C) 180 days   D) 360 days

41. On a commercial vehicle utilized in the practice of electrical contracting the required identification information shall visible mark on the vehicle with lettering at least_____ height.

A) 2 in   B) 3 in   C) 4 in   D) 5 in

42. A corporation doing business in NJ that is required to pay estimated taxes will pay ___ of the current estimated tax liability in the 6th month of the accounting period.

A) 25%   B) 30%   C) 50%   D) 60%

43. Which of the following is NOT governed by the rules and regulations of the State Board Of Examiners.

A) Your pricing B) Your continuing education C) Workers supervision D) Marketing practices

44. _____ is a bond protecting the owner from loses associated from a successful bidder not honoring the bid.

A) Bid Bond   B) Holding Bond   C) Performance Bond   D) Lien Bond

45. How much sales tax must be collected on a bill totaling $650.00 for a service panel replacement where the bill states in writing charges billed 1. Material : $200 ; 2. Labor: $450 ?

   A) $12   B) $27   C) $39   D) none required

46. Employers with more than ____ employees must also maintain records of occupational injuries.

   A) 6   B) 10   C) 11   D) 12

47. The processes of settling contract disputes without costly judicial trails are known as _____.

   A) Litigation
   B) Contract Litigation Resolution
   C) Alternative Contract litigation
   D) Alternative Dispute Resolution

48. A _____ of the tax liability penalty is imposed on an employer for "trust fund" tax payments, defined as withheld taxes, paid 10 days late.

   A) 2%   B) 100%   C) 10%   D) 5%

49. Electrical Contractors are required to contact the _____ prior to any excavation to request marking of underground utilities.

   A) Common Ground Alliance
   B) 1-800-Mark
   C) Electric Utility Co.
   D) NJ One Mark

50. If the liquidity ratio for a company is greater than one the company is _____.

   A) in a poor position to pay its current debts
   B) in a positive position to pay its current debts
   C) will be unable to pay its debts
   D) facing a negative cash flow

Chapter 15 Chapter Review and Exam Answers

This chapter contains all the answers to the chapter review and practice exams with references back to the section within the NJ Electrician's reference manual. The answers are listed as followers;

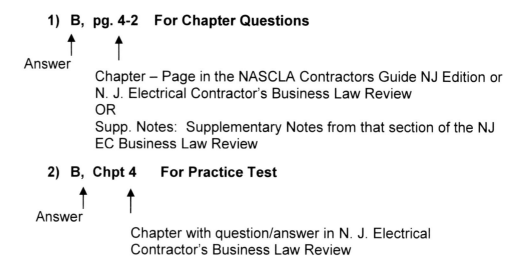

1) B, pg. 4-2   For Chapter Questions

Answer

Chapter – Page in the NASCLA Contractors Guide NJ Edition or N. J. Electrical Contractor's Business Law Review
OR
Supp. Notes: Supplementary Notes from that section of the NJ EC Business Law Review

2) B, Chpt 4   For Practice Test

Answer

Chapter with question/answer in N. J. Electrical Contractor's Business Law Review

Chapter 1
1) C, pg. 1-2
2) A, pg. 1-2
3) B, pg. 1-2
4) B, pg. 1-2
5) D, pg. 1-2
6) C, pg. 1-2

Chapter 2
1) D, pg. 2-1
2) A, pg. 2-1
3) B, pg. 2-1
4) B, pd. 2-2
5) A, pg. 2-2
6) A, pg. 2-2
7) C, pg. 2-2
8) D, Pg. 2-2
9) A, pg. 2-3
10) C, pg. 2-3
11) D, pg. 2-3
12) C, pg. 2-5
13) C, pg. 2-3, 2-4
14) B, pg. 2-4
15) B, pg. 2-4
16) D, pg. 2-5
17) D, pg. 2-5
18) B, Appx A
19) A, Appx A
20) D, pg. 2-6
21) B, pg. 2-6
22) A, pg. 2-6
23) D, pg. 2-6
24) B, pg. 2-7

Chapter 3
1) B, pg. E-2, 45:5A-3
2) A, pg. E-2, 45:5A-3
3) D, pg. E-2, 45:5A-3
4) B, pg. E-2, 45:5A-3
5) B, pg. E-4, 45:5A-9.1
6) D, pg. E-4, 45:5A-9.1
7) A, pg. E-6, 45:5A-13
8) B, Chpt. 3 Supplementary Notes
9) D, pg. E-7, 45:5A-13.3
10) A, pg . E-7, 45:5A-13-3
11) B, pg. E-7, 45:5A-13.6
12) A, pg. E-8, 45:5A-14
13) A, pg. E-8/E-11, 45:5A-18/13:31-1.2
14) D, pg. E-8/E-11, 45:5A-18/13:31-1.2

15) B, pg. E-11, 45:5A-19
16) B, pg. E-12, 13:31-1.5
17) D, E-12, 13:31-1.5
18) A, E-12, 13:31-1.5 b
19) A, E-13,13:31-1.5 e & 45:5A
20) C, E-13, 13:31-1.7
21) B, E-13, 13:31-1.7
22) B, E-14c, 13:31-1.7
23) A, E-14c, 13:31-1.7
24) A, E-14d, 13:31-1.7
25) C, E-14e, 13:31-1.7
26) C, E-15g, 13:31-1.7
27) B, E-16a, 13:31-1.8
28) D, E-17d, 13:31-1.8
29) B, E-18c1, 13:31-2.2
30) B, E-18c2, 13:31-2.2
31) A, E-18c3, 13:31-2-2
32) D, E-18e, 13:31-2.3
33) D, E-22, 13:31-3.1
34) D, E-23g, 13:31-3.3
35) A, E-24d, 13:31-3.4
36) C, E-24d, 13:31-3.4
37) C, E-24c, 13:31-3.4
38) B, E-24a, 13:31-3.5
39) D, E-25a, 13:31-3.7
40) A, E-26 , 13:31-4.1
41) C, E-26c4, 13:31-4.1
42) B, E-27a1, 13:31-4.2
43) D, E-28, 13:31-4.3
44) D, E-28, 13:31-5.1
45) C, E-32, 13:31-5.1
46) D, H-22, 13:31A-3.1
47) C, H-26, 13:31A-3.6
48) B, H-26, 13:31A-3.7
49) C, I-7, 45:1-11
50) C, I-12, 45:1-18.2

Chapter 4
1) B, pg. 4-1
2) A, Supp Notes
3) B, pg. 4-3
4) B, Supp Notes
5) C, pg. 4-3
6) C, pg. 4-3
7) B, Supp Notes
8) C, pg. 4-4
9) D, pg. 4-4
10) B, pg. 4-4
11) A, pg. 4-4
12) D, Supp Notes

13) B, pg. 4-3
14) C, pg. 4-4
15) B, pg. 4-4
16) A, pg. 4-6
17) A, pg. 4-6, 15-4
18) B, pg. 4-5
19) D, pg. 4-6
20) A, pg , 4-2,5,6
21) C, Supp Notes
22) D, pg. 4-7
23) C, Supp Notes
24) A, pg. 4-7
25) C, pg. 4-7
26) D, Supp Notes
27) D, pg. 4-8
28) A, pg. 4-8
29) A, pg 4-8
30) A, pg. 4-8
31) B, pg. 4-9
32) B, pg. 4-9
33) A, pg. 4-9
34) C, pg. 4-9
35) B, pg. 4-9

Chapter 5

1) A, pg. 7-1
2) B, pg. 7-1
3) D, pg. 7-1
4) C, pg. 7-1
5) B, pg. A-1
6) A, pg. 7-2
7) B, pg. 7-2
8) D, pg. 7-2
9) C, pg. A-1
10) D, pg. 7-1
11) B, Supp Notes
12) C, pg.7-2
13) D, Supp Notes
14) A, pg. 7-3
15) A, pg. 7-3
16) A, pg. 7-4
17) C, pg. 7-4 & Supp Notes
18) A, pg. 7-4 & 7-6
19) C, pg. 7-4
20) A, pg. 7-6
21) B, pg. 7-5
22) C, pg. 7-6 10% of $5500=$550 added back
    to the $5500 for a Bid price of $6050
23) B, pg.7-7
24) D, pg. 7-7

Chapter 6

1) A, pg. 8-1
2) D, pg. 8-1/2 (not legal scope)
3) C, pg. 8-1
4) C, pg. 8-1
5) A, pg. 8-2
6) B, pg. 8-2
7) B, pg. 8-3
8) A, pg. 8-3 & 8-4 Standard term used for
   section defining obligation of the parties
9) D, pg. 8-4
10) C, pg. 8-4
11) C, pg. 8-4
12) B , pg. 8-4/A-7
13) A, pg. 8-4/A-6
14) B, pg. 8-4
15) D, pg. 8-5
16) D, pg. 8-5
17) A, pg. 8-5
18) A, pg. Supp. Notes
19) C, pg. 8-5
20) C, pg. 8-5
21) C, pg. 8-6
22) B, pg. 8-6
23) A, pg. 8-5
24) B, pg. 8-6
25) D, pg. 8-6
26) B, pg. 8-6
27) A, pg. 8-7
28) C, pg. 8-7
29) D, pg. 8-7
30) B, pg. 8-8
31) A, pg. 8-8
32) C, pg. 8-8
33) D, pg. 8-8

Chapter 7

1) D, pg. Supp. Notes
2) C, pg. 9-2
3) C, pg. 9-2
4) B, pg. 9-2
5) B, pg. 9-2
6) A, pg. 9-2
7) C, pg. 9-3/4
8) A, pg. 9-5
9) D, pg. 9-5
10) A, pg. 9-6
11) B, pg. 9-6
12) A, pg. 9-7

13) B, pg. 9-6
14) C, pg. 9-7
15) D, pg. 9-7
16) B, pg. 9-9
17) B, pg. 9-9
18) C, pg. 9-10
19) D, pg. 9-10
20) B, pg. 9-10
21) C, Supp. Notes
   88 outlets/(3 men/hr * 8 hr/day * 3 days)=1.22

31) C, pg. 11-12
32) A, pg. 11-12
33) B, pg. 11-12
34) B, pg. 11-12
35) D, pg. 11-12/13
36) B, pg. 11-13
37) D, pg. 11-13
38) A, pg. 11-13
39) C, pg. 11-13
40) C, pg. 11-13
41) A, pg. 11-14

Chapter 8
1) C, pg. 11-1
2) A, pg. 11-1
3) D, pg. 11-2
4) B, pg. 11-2
5) C, pg. 11-2
6) A, pg. 11-2
7) A, pg. 11-2
8) D, pg. 11-3
9) D, pg. 11-3
10) A, pg. 11-3
11) B, pg. 11-3
12) C, pg. 11-4
13) C, pg. 11-4 & Supp. Notes
   $10 * 40 hrs   = $400
   $10 * 1.5 (time and ½) * 10 hrs = $150
                              Total Gross = $550
14) C, pg. 11-4
15) C, pg. 11-4
   $10 * 40 hrs   = $400
   $10 * 1.5 (time and ½) * 10 hrs = $150
   $10 * 20 hrs   = $200
                              Total Gross = $750
16) A, pg. 11-4
17) D, pg. 11-5
18) B, pg. 11-5
19) C, pg. 11-5
20) A, pg. 11-5
21) C, pg. 11-6
22) D, pg. 11-6
23) D, pg. Supp. Notes
24) A, pg. 11-8
25) B, pg. 11-8
26) C, pg. 11-8
27) A, pg. 11-8
28) D, pg. 11-8
29) A, pg. 11-8
30) B, pg. 11-9

Chapter 9
1) B, pg.12-1
2) C, Supp. Notes
3) B, Supp. Notes
4) B, Supp. Notes
5) C, pg. 12-2
6) B, pg. 12-3
7) B, pg. 12-3
8) A, pg. 12-3
9) A, pg. 12-3/4
10) B, pg. 12-4
11) D, pg. 12-4
12) C, pg. 12-4
13) C, pg. 12-4
14) D, pg. 12-4
15) D, pg. 12-4
16) B, pg. 12-4
17) B, pg. 12-6
18) D, pg. 12-6
19) A, pg. 12-6
20) B, pg. 12-6
21) B, pg. 12-7
22) C, pg. 12-8
23) A, pg. 12-8
24) C, pg. 12-8
25) D, pg. 12-9
26) A, pg. 12-9
27) A, pg. 12-9
28) C, pg. 12-10
29) A, pg. 12-10
30) D, pg. 12-10
31) C, pg. 12-11
32) B, pg. 12-11
33) A, pg. 12-12
34) D, pg. 12-14

Chapter 10
1) B, pg. 14-2
2) C, pg. 14-2
3) D, pg. 14-2
4) D, pg. 14-2
5) A, pg. 14-2
6) B, pg. 14-2
7) A, pg. 14-2
8) C, pg. 14-2
9) B, pg. 14-2
10) D, pd. 14-3
11) C, pg. 14-3
12) A, pg. 14-3
13) B, pg. 14-3/4
   $26,400 - $8,200 = $18,200
14) B, pg. 14-3
15) A, pg. 14-3
16) A, pg. 14-3
17) C, pg. 14-5
18) A, pg. 14-5
19) D, pg. 14-5
20) C, pg. 14-5
21) B, pg. 14-5
22) D, pg. 14-5
23) D, pg. 14-7
24) A, pg. 14-7
25) C. pg. 14-7
26) A, pg. 14-7
27) B, pg. 14-7
28) D, pg. 14-7
29) C, pg. 14-7
30) C, pg. 14-8
31) D, pg. 14-8
32) B, pg. 14-8
33) A, pg. 14-8
34) A, pg. 14-8
35) B, pg. 14-8
36) A, pg. 14-9
37) C, pg. 14-9
38) D, pg. 14-9
39) B, pg. 14-9
40) B, pg. 14-10
41) A, pg. 14-10
42) D, pg. 14-10
43) C, pg. 14-10
44) D, pg. 14-11
45) B, pg. 14-11
46) C, pg. 14-11
47) B, pg. 14-11
48) B, pg. 14-11

49) A, pg. 14-11
50) C, pg. 14-11
51) A, pg. 14-12
   $2000/5 = $400 / yr
   3 * $400/yr = $1200 depreciation in three years
52) D, pg. 14-12
53) A, pg. 14-12
54) C, pg. 14-13 3% discount if paid by 1/12/1
55) D, pg. 14-13 3% discount if paid by 2/10/1
56) A, pg. 14-13
57) C, pg. 14-15
58) B, pg. 14-15
59) D, pg. Chpt 14 Suppl. Notes
60) A, pg. Chpt 14 Suppl. Notes
61) C, pg. Chpt 14 Suppl. Notes
62) D. pg. 14-14
63) C , pg. Chpt 14 Suppl. Notes
   Withheld Tax: SS $62
           Medicare $14.50
   Employer: SS $62
           Medicare $14.50
           FUTA $62
   Total Tax Liability: $215
64) D, pg. Chpt. 14 Suppl. Notes

Chapter 11
1) A, pg. 15-1
2) C, pg. 15-1
3) D, pg. 15-1
4) B, pg. 15-3
5) D, pg. 15-3
6) B, pg. 15-3
7) C, pg. 15-3
8) B, pg. 15-4
9) B, pg. 15-4
10) C, pg. 15-4
11) C, pg. 15-4/Supp. Notes
12) A, pg. Supp. Notes
13) D, pg. Supp. Notes
14) A, pg. 15-4
15) A, pg. 15-4
16) B, pg. 15-4
17) A, pg. 15-4
18) D, pg. 15-4
19) D, pg. 15-5
20) A, pg. Supp. Notes
21) C, pg. 15-5
22) A, pg. 15-5
23) A, pg. 15-5

24) B, pg. 15-5
25) D, pg. 15-5
26) C, pg. 15-5
27) B, pg. 15-5
28) B, pg. 15-6
29) C, pg. 15-6
30) A, pg. 15-7
31) D, pg. 15-7
32) C, pg. 15-7
33) A, pg. 15-7
34) D, pg. 15-7
35) B, pg. 15-7
36) C, pg 15-8 & Supp. Notes
37) B, Supp. Notes (Labor only)
38) B, pg. 15-8

Chapter 12
1) C, pg. 16-1
2) A, pg. 16-2
3) B, pg. J-8 2A:44A-8
4) A, pg. J-6 2A:44A-7
5) A, pg. 16-2/J-15 2A:44A-20
6) B, pg. J-14 2A:44A-14
7) C, pg. J-13 2A:44A-14

Chapter 13
1) C, pg. 5-2
2) A, pg. 5-2
3) C, pg. 5-2
4) D, pg. 5-2
5) B, pg. 6-1, A-7
6) B, pg. A-7
7) D, pg. 10-2
8) D, pg. 13-2/3
9) B, pg. 13-3

Practice Exam #1
1) A, pg. 1-2
2) D, pg. 2-1
3) A, pg. 2-1
4) D, pg. 1-2
5) A, pg. 2-4
6) A, pg. E-8/E-11, 45:5A-18/13:31-1.2
7) A, pg. 2-6
8) D, pg. E-2, 45:5A-3
9) B, pg. E-2, 45:5A-3
10) B, pg. E-12, 13:31-1.5
11) B, pg. 4-3

12) C, pg. 7-1
13) D, E-12, 13:31-1.5
14) A, pg. 8-1
15) A, E-14c, 13:31-1.7
16) A, E-24d, 13:31-3.4
17) B, pg. 8-3
18) C, I-7, 45:1-11
19) A, pg. 8-4/A-6
20) A, E-18c3, 13:31-2-2
21) C, pg. 9-2
22) D, pg. 11-2
23) B, pg. 11-12
24) B, E-27a1, 13:31-4.2
25) B, pg.12-1
26) B, pg. 9-6
27) A, pg. 12-3
28) C, Supp Notes
29) C, pg. 14-2
30) B, pg. 14-9
31) A, pg. 4-9
32) C, pg. 15-1
33) D, pg. 8-6
34) A, pg. 14-12
    $2000/5 = $400 / yr
    3 * $400/yr = $1200 depreciation in three years
35) D, Supp Notes
36) A, pg. 15-4
37) A, pg. 16-2
38) A, pg. 7-4
39) C, pg. 15-4/Supp. Notes
40) C, pg. 7-6 10% of $5500=$550 added back to the $5500 for a Bid price of $6050
41) B, pg. 14-8
42) A, pg. J-6 2A:44A-7
43) A, pg. 4-6, 15-4
44) C, pg. 9-7
45) D, pg. 7-1
46) D, pg. 8-8
47) C, Supp. Notes
    88 outlets/(3 men/hr * 8 hr/day * 3 days)=1.22
48) C, pg. 14-8
49) C, pg. 9-10
50) A, pg. 11-2

Practice Exam #2

1) B, 2A:44A-30
2) D, pg. 14-2
3) C, pg. 11-2
4) D, pg. 9-10
5) B, pg. 14-10
6) C, pg. 8-8
7) A, pg. 7-6
8) A, pg. 14-9
9) A, pg. 4-9
10) D, pg. 1-2
11) A, pg. 7-3
12) D, pg. 9-7
13) B, pg. 2-4
14) C, pg. 2-2
15) D, pg. 15-3
16) B, pg. 7-2
17) C, pg. 9-3/4
18) A, pg. 14-2
19) C, pg. 1-2
20) A, pg. 2-6
21) B, pg. 15-4
22) B, pg. 8-4
23) B, E-18c1, 13:31-2.2
24) A, pg. 11-5
25) A, E-14d, 13:31-1.7
26) A, pg. 4-8
27) A, pg. 15-4
28) C, I-12, 45:1-18.2
29) C, pg. 4-4
30) B, pg. 2-7
31) D, E-28, 13:31-5.1
32) C, E-24d, 13:31-3.4
33) B, pg. 8-6
34) B, pg. 14-3
35) C, pg. 12-2
36) B, pg. 11-12
37) D, pg. 4-6
38) A, pg . E-7, 45:5A-13-3
39) B, pg. 14-5
40) D, E-23g, 13:31-3.3
41) B, pg. E-12, 13:31-1.5
42) C, pg. 15-7
43) D, E-25a, 13:31-3.7
44) A, pg. 4-7
45) B, Supp. Notes (Labor only)
46) B, pg. 12-3
47) D, pg. 8-7
48) D, pg. 15-5
49) A, pg. 12-6
50) B, pg. 14-7